BUILDING NUMBER SENSE with TEN-FRAMES

This book belongs to:

number +

www.numberplus.co.uk

First published 2026
001

Illustrations by Marco Zerneri

paula.claughton@numberplus.co.uk

Scan the QR code
to download bonus worksheets

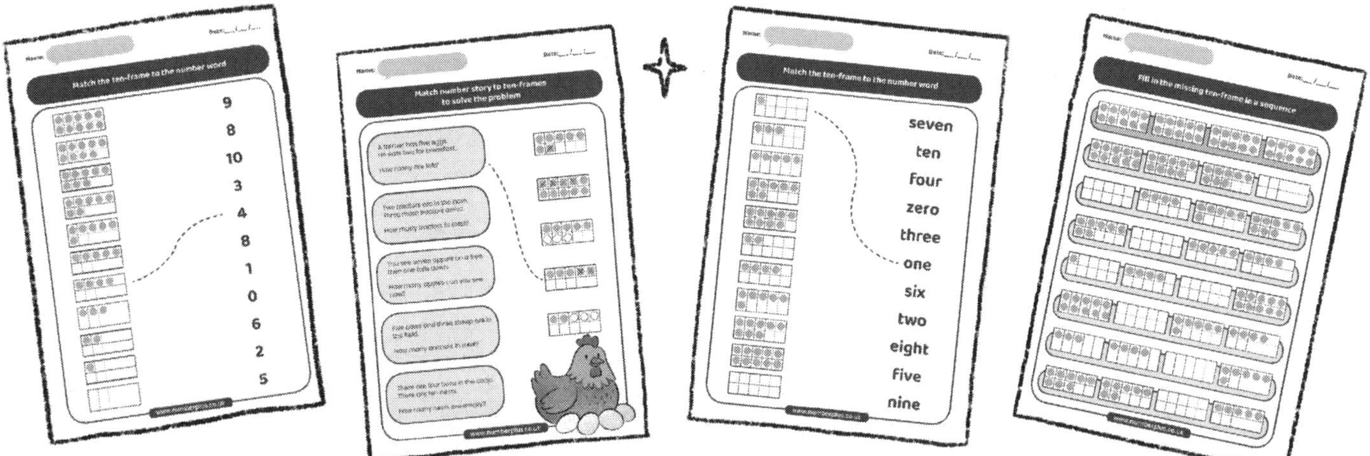

Contents

What is a ten-frame?

A ten-frame is a simple maths tool that helps young children **see** and **understand** numbers.

It looks like a rectangle with 10 boxes - two rows of five.

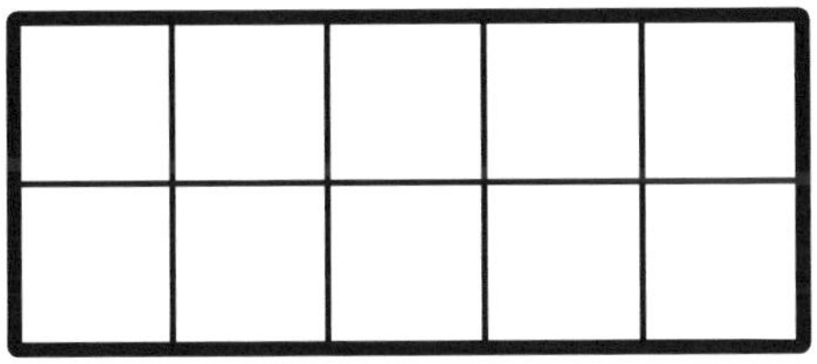

It helps children quickly see how many items are in a group without having to count them one by one.

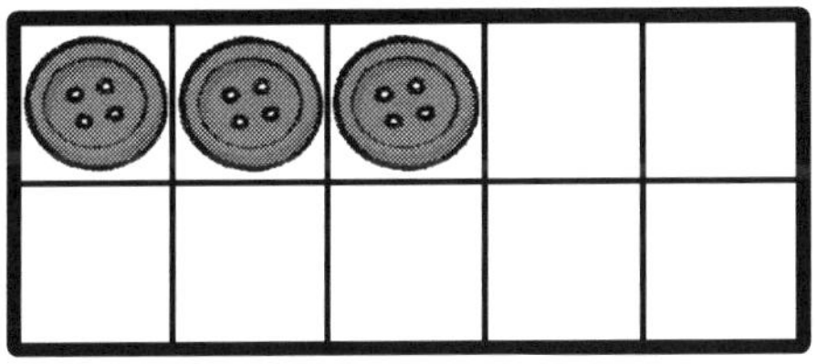

Example: *3*

It shows missing pieces. If some boxes are empty, children can figure out how many more you need to make 10.

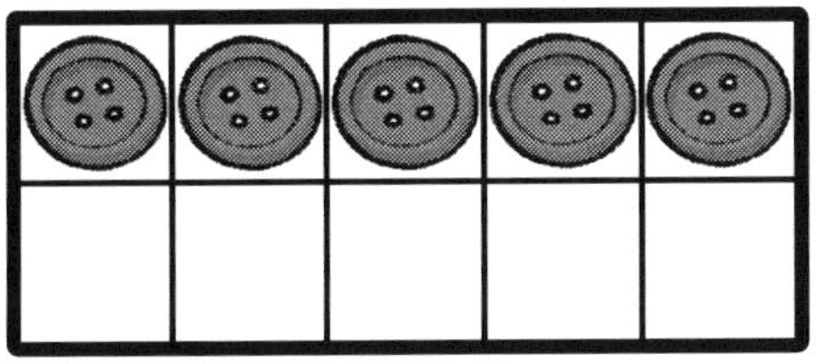

Example: *5 + ☐ = 10*

Why is a ten-frame helpful?

Ten-frames are great for young learners because they:

- Help children spot patterns, like double 5 makes 10
- Make it easier to calculate
- Build a strong sense of how numbers work
- Help children recognise numbers by sight

How does this book help my child?

We use a simple three-step method to help your child understand numbers deeply.
It's called **Concrete–Pictorial–Abstract (CPA)**:

- **Concrete** Children play with real objects like blocks or buttons to show numbers.
- **Pictorial** They use pictures or drawings to see the numbers in a new way.
- **Abstract** They start to use number symbols (like 5 or +) to solve problems.

This step-by-step approach helps children move from playing with objects to thinking with numbers.

To make things easier for young children, we've used simple, child-friendly words instead of tricky maths terms.

This key shows you the words we've chosen to help your child understand as they learn:

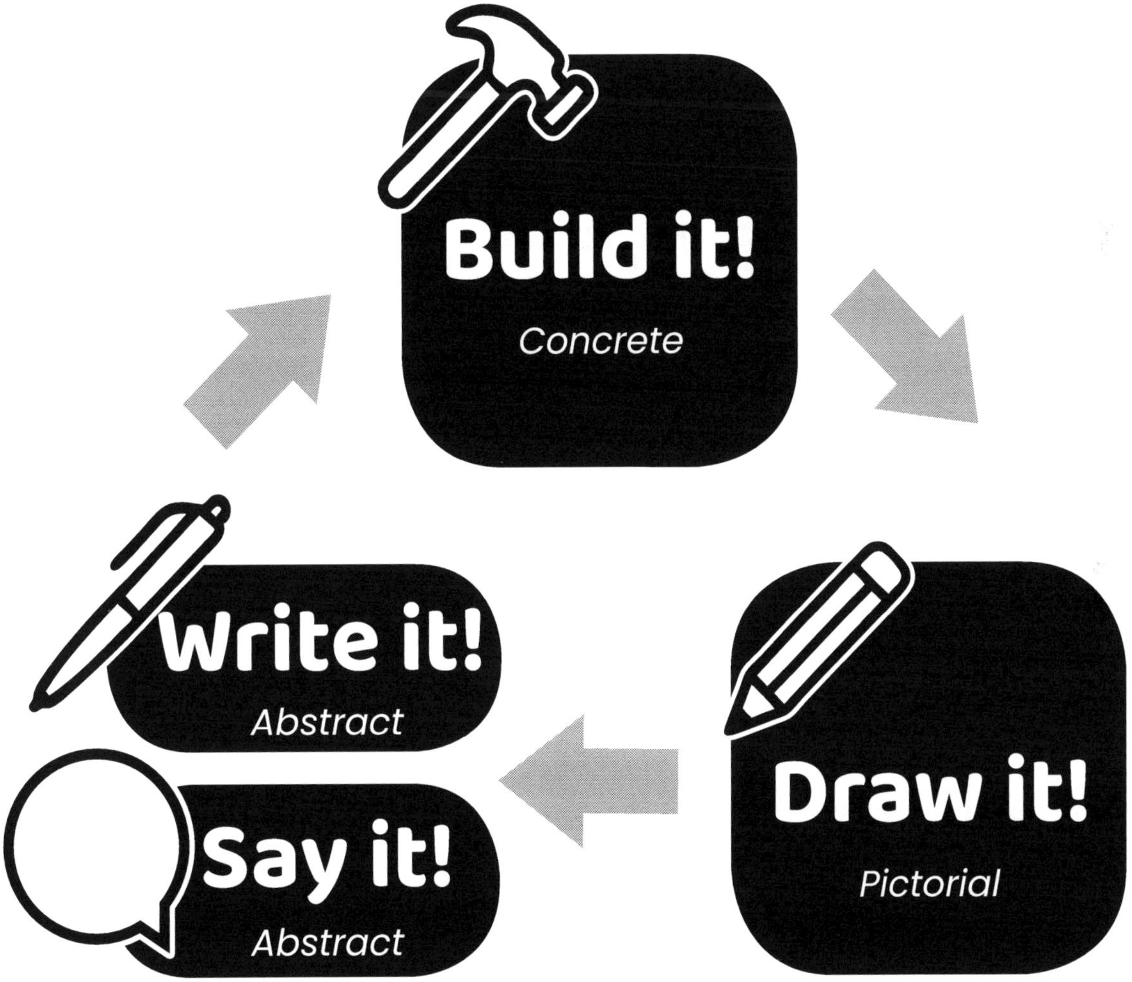

Spotting Number Patterns

Ten-frames help children build early number skills by showing numbers in simple patterns. Using different patterns on a ten-frame helps children build number sense and notice important relationships. *For example*:

2s Pattern

Place counters in pairs—one on top, one below.

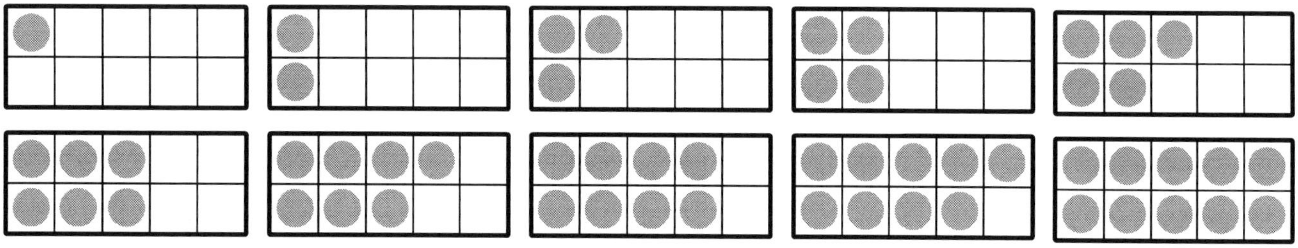

This helps children spot even and odd numbers and practise counting in twos.

5s Pattern

Place objects across the top row from left to right, then continue on the bottom row.

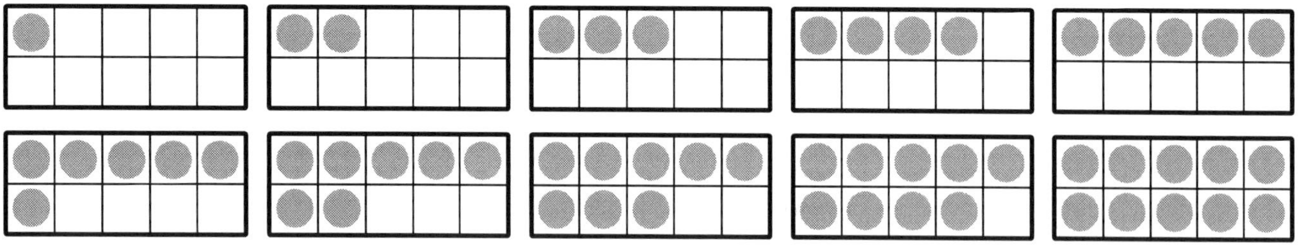

This helps children see how numbers are grouped in 5s and 10s.

There's no one "right" way to use a ten-frame, what matters is helping children make sense of numbers. Encourage them to try both patterns and notice what works best for them.

Finger Patterns

Young children love using their fingers, and they're a brilliant maths tool!

0

Finger patterns help children:

- See numbers quickly without counting
- Understand number combinations (*like 6 and 4 make 10*)
- Build confidence and fine motor skills

1 2 3 4 5

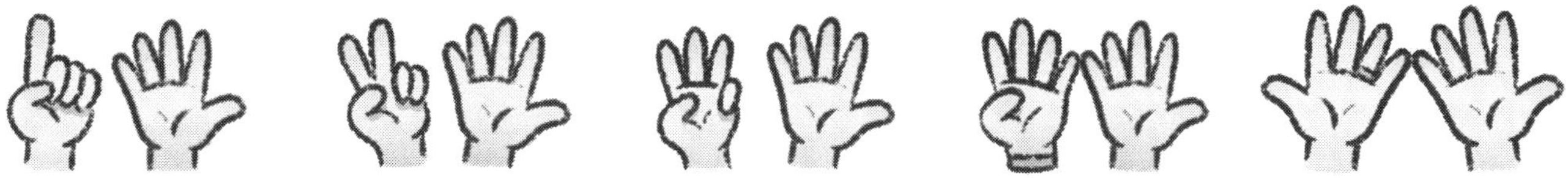

6 7 8 9 10

Look for this icon in activities.

It's a friendly reminder to use fingers alongside the ten-frame!

What do you notice?

This book gently changes how problems are shown, so your child can spot patterns and think flexibly.
For example, children might:

Build 7 on a ten-frame using objects

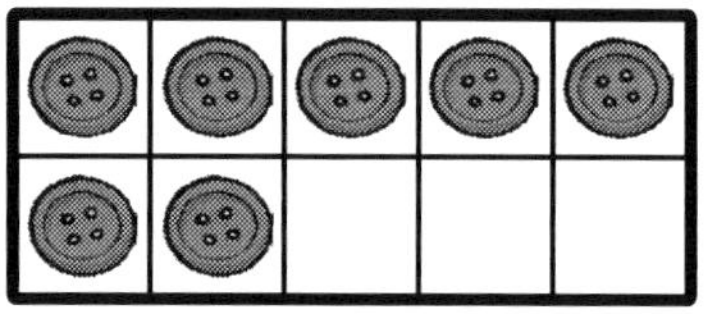

Draw 7 dots on a ten-frame

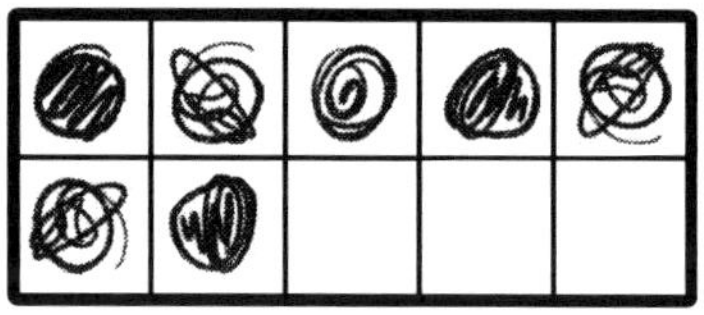

Show 7 as a finger pattern

Also, by carefully changing the steps, children start to notice what stays the same and what's different, helping them make sense of how maths works.

Example:

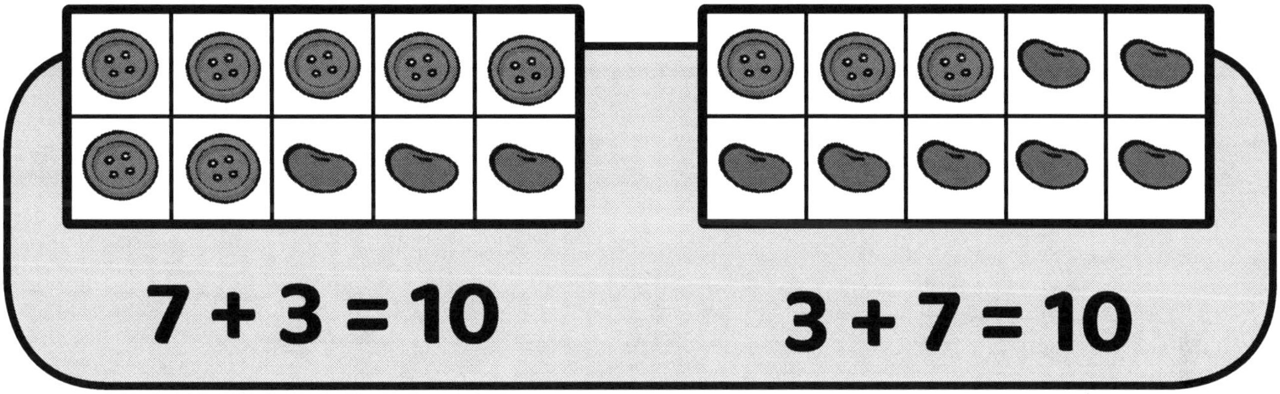

?

What's the same?
Both make 10.

?

What's different?
The numbers are in a different order.

This helps build number sense, an understanding of how numbers work and relate to each other.

Simply put: It's like learning with Lego — children see how the pieces fit together so they can build anything on their own.

To find out more, please go to the 'Answers Explored' section at the back of this book

Let's talk about it!

Talking about maths helps your child understand and remember what they learn.

When we explain our thinking, our brains get stronger, and we learn new ways to solve problems.

Ten-frames help children talk about maths by giving them a clear way to see and explain numbers, patterns and relationships, such as:

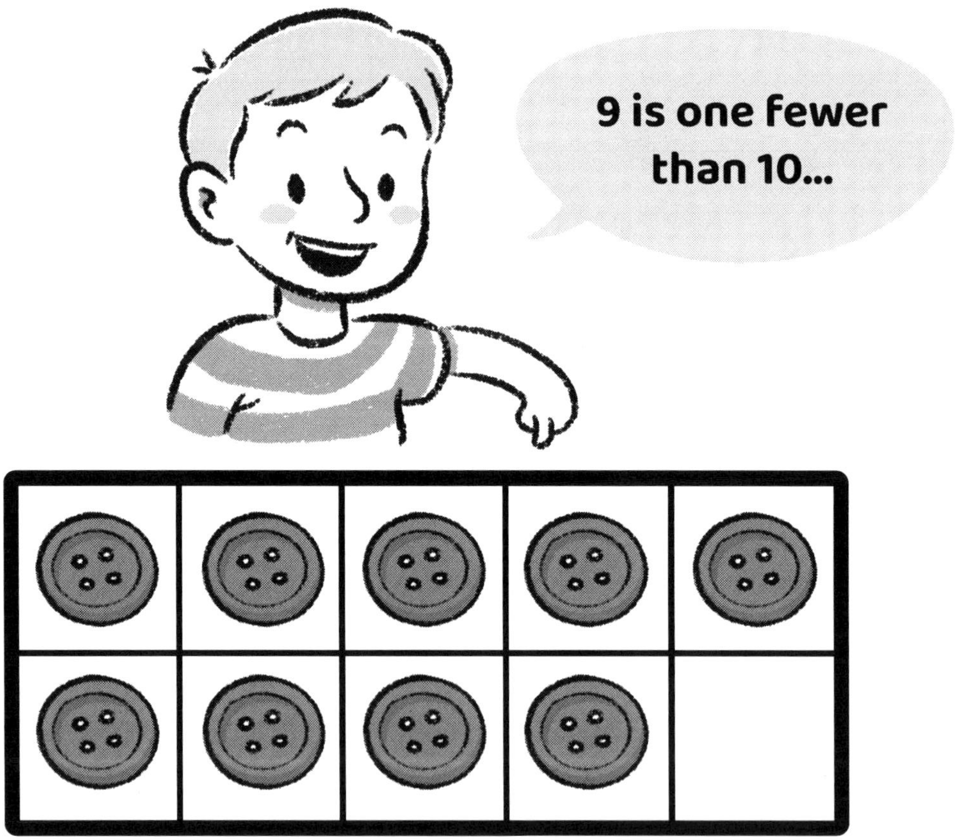

You can help by asking questions like:

What do you see?

What's the same?

How did you work that out?

What's different?

What happens if we add one more?

These **little** chats help children grow **big** maths ideas!

Helpful Hints

- Sit together and enjoy the number stories.
- Use small objects (buttons, cereal) as counters.
- Ask simple questions like 'Can you show me 5?' or 'How many do you have?'
- Let them notice patterns like rows of 5 on a full ten-frame.
- It's okay if they make mistakes, mistakes help us learn!
- Cheer them on. Encouragement goes a long way.

Getting started

Turning everyday objects like empty egg boxes, buttons, or snacks into maths tools helps children make connections between numbers and the real world. It also brings maths to life in a fun and meaningful way so, let's get started!

First, cut out the ten-frame templates from the back of this book OR...build your own using everyday objects.
Here are a few ideas to build ten-frames with children...

Empty box lid

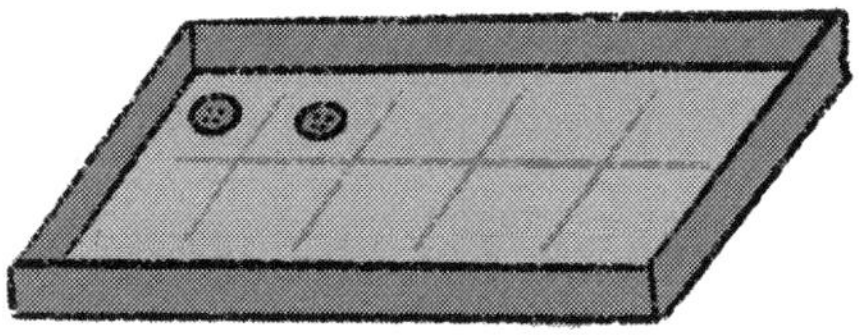

Sticks

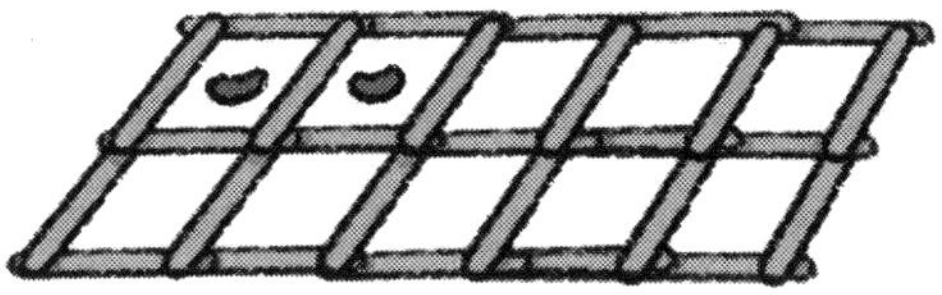

Ice cube tray

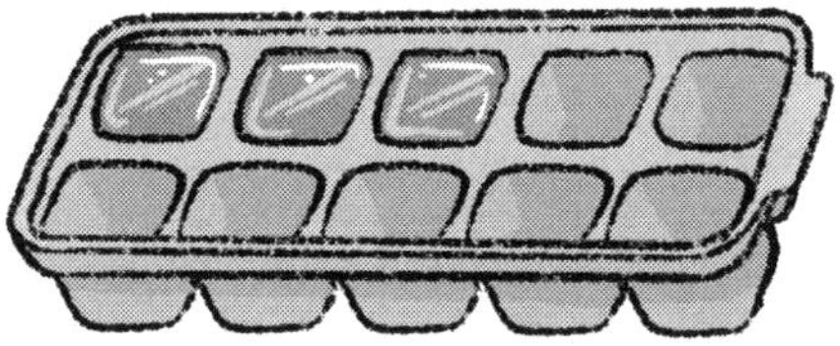

Egg boxes

Chalk

Tape

Then, cut out the counters at the back of this book OR...use some small objects which are the same or similar size, shape and colour, like...

Buttons

Coins

Small Toys

Dried pasta

Pebbles

Shells

Now, have fun playing the **'Build It!'** activities with young learners.

Top Tip: Laminate or glue resources onto card so they last longer.

Build it!

Have ready:

- 2 empty ten-frames
- objects or counters (20)
- number cards (0-20)
- flashcards (0-10)

Optional:

- building blocks (such as, Lego, Mega Bloks, wooden cubes)

Build it!

This step supports the **concrete** stage of learning, which is essential at this age. Using real objects helps children build strong mental images of numbers they can picture and remember later.

Your child might:
Fill one row before starting the next (to show 5s and 10s),

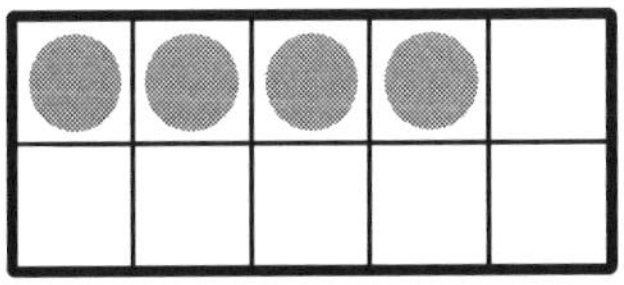

Make vertical pairs (to show 2s and even numbers),

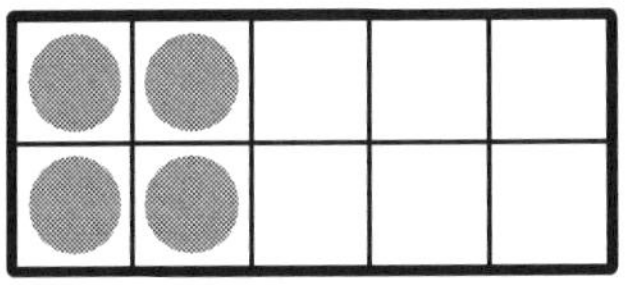

Or use shapes and colours to build their own patterns.

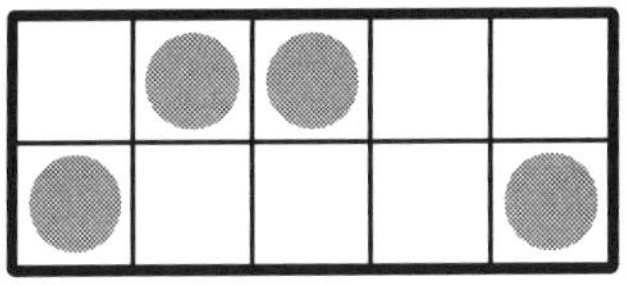

Important Point
While exploring is great, try encouraging structured ones (like filling from the top left) so number patterns are clearer and easier to talk about.

We've included 12 fun activities to help your child get used to using ten-frames. These games are a super way to help them see, touch, and build numbers, right from the start!

1. Ten-Frame Explorer

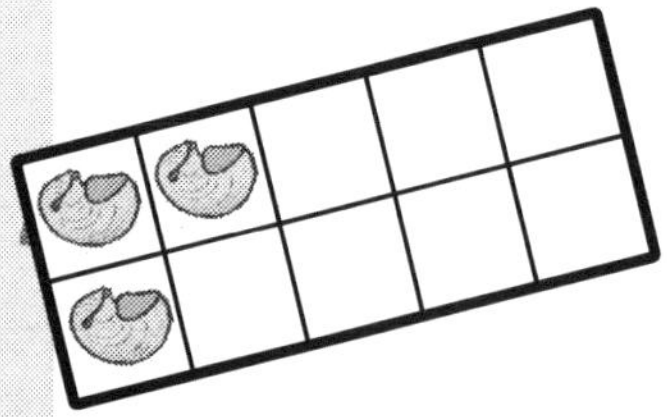

Let children explore a ten-frame by filling it with counters in any way they like. Discuss how many counters they used and how many spaces are left.

This activity familiarises children with the structure of the ten-frame.

2. Number Card Match

Shuffle the number cards and place face down in a pile. Turn over the first card and place that many counters on a ten-frame. Check if the counters match the number on the card. Repeat.

This activity helps children practice one-to-one correspondence and number recognition.

3. Guess the number

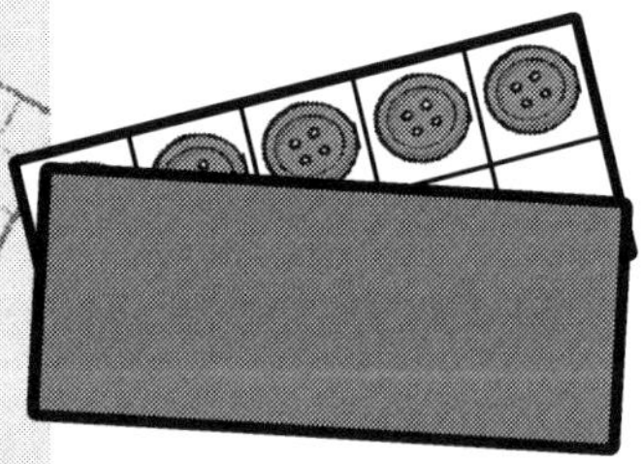

Secretly place some counters on a ten-frame and cover it. Show briefly and ask the child to guess how many without counting.

This activity helps children practice subitising and estimating.

4. Build and Compare

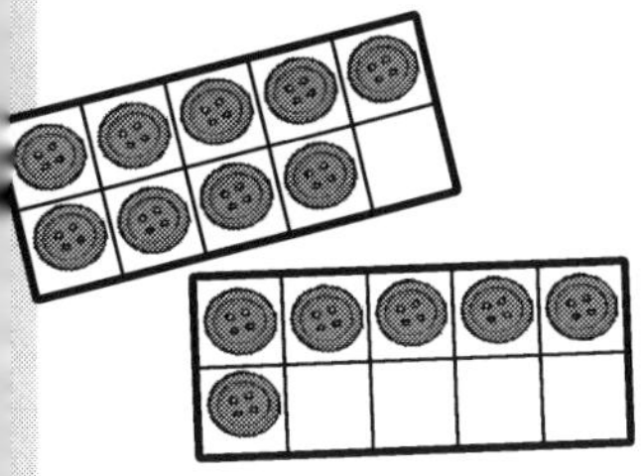

Shuffle the number cards and place face down in a pile. Turn over two number cards and build both numbers on separate ten-frames. Compare which has more or fewer counters.

This activity helps children compare quantities and understand greater/less than.

5. Ten-Frame Towers

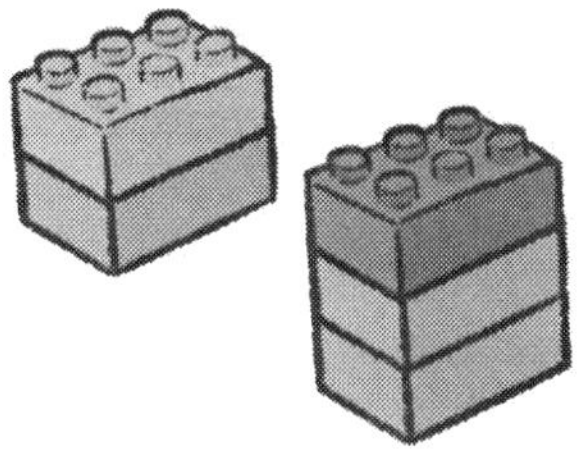

Build a cube tower matching the number of counters on a ten-frame. Compare towers to see which is taller or shorter.

This activity helps children to visually link numbers with quantity.

6. Missing Counters

Shuffle the number cards and place face down in a pile. Turn over a number card and place that many counters on a ten-frame. Ask, "How many more to make ten?"

This activity helps children develop an understanding of number bonds to ten.

7. Double Trouble

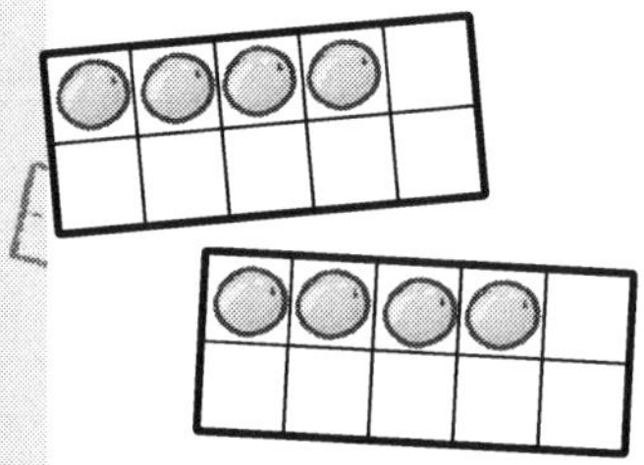

Shuffle the number cards and place face down in a pile. Turn over a number card and place that many counters on a ten-frame. Build the same number on another ten-frame and find the double.

This activity introduces doubling and addition.

8. Memory Match

Create a matching game with number cards (0-10) and flashcards. Turn over two at a time to find pairs. Example: Turn over 3 and a matching frame – keep the pair if correct.

This activity strengthens memory and reinforces the connection between numbers and visual models.

9. More or Fewer

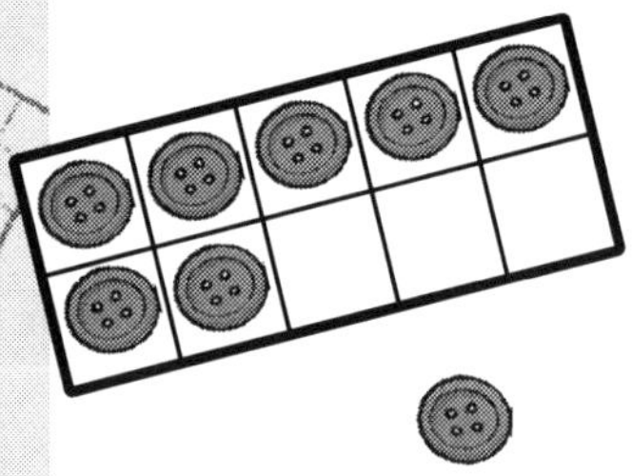

Shuffle the number cards and place face down in a pile. Turn over a number card and place that many counters on a ten-frame. Then add one more or take one away and say the new number. For extra challenge, add two more or take two away.

This activity helps children develop addition and subtraction skills with small numbers.

10. Once Upon a Time...

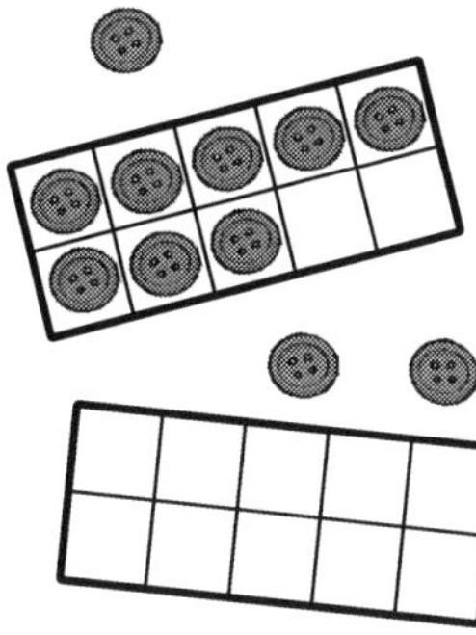

Create a simple story where characters arrive or leave a group. (For example, *'Once upon a time there were 8 frogs in a pond. 3 more jumped in. How many now?'*). Build the story using two ten-frames to find the total.

This activity helps children to connect number concepts with storytelling.

11. Ten-Frame Addition

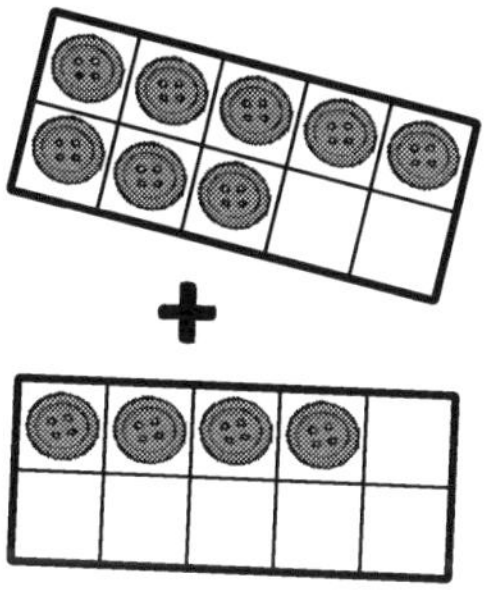

Shuffle the number cards and place face down in a pile. Turn over two number cards and build both numbers on separate ten frames. Combine the counters to find the total.

This activity helps children understand basic addition within 20.

12. Fill the Frames

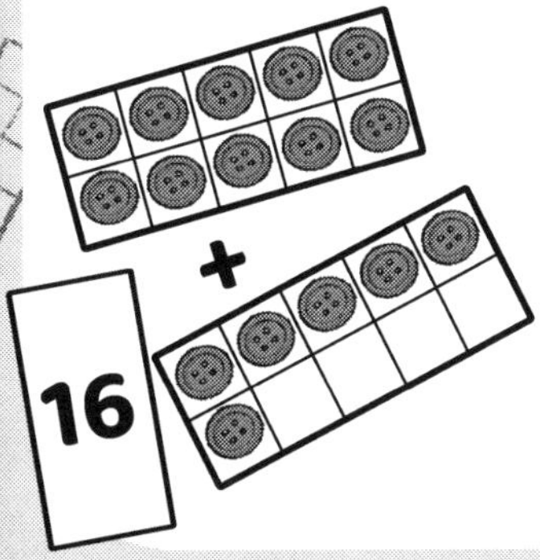

Shuffle number cards (10-20) and place face down on a flat surface. Turn over the first card (e.g., 16). Child fills ten-frames with the correct number of counters (1 full frame and 6 on the next). Ask: 'What do you see?' (10 and 6, that makes 16!). Repeat with a different 2-digit number.

This activity helps children compose numbers using tens and ones.

Draw it!

Have ready:

- 2 ten-frames
- objects or counters (20)
- something to write with

Top Tip!
Use words like *add, plus, more, take away, minus, subtract, altogether, total, how many more*, and *how many left* as your child draws and explains.

Draw it!

This step is all about visualising numbers—it's part of the **pictorial** stage of learning. Children move from using real objects (concrete) to drawing what they see — a key step from doing maths to thinking in pictures. This bridges hands-on experiences and symbolic thinking.

Simple number stories in this section make maths meaningful (and fun).

Example: There were 8 ducks in a pond. Then 4 more waddled in. How many ducks now? Encourage your child to draw 8 dots, then add 4 more.

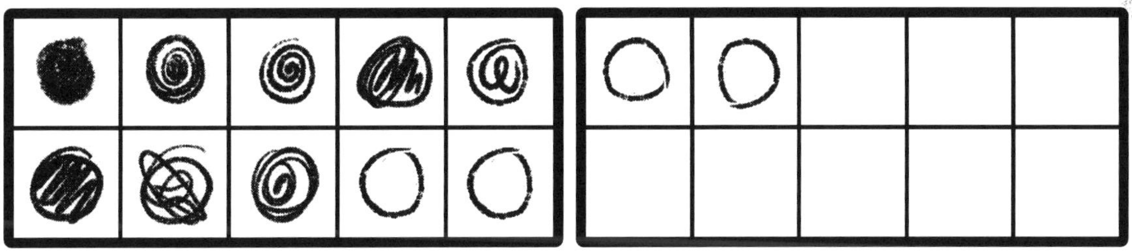

Ten-frames help children see how numbers are built from tens and ones, showing patterns that make counting and adding easier. Over time, they can picture numbers mentally even without objects.

Just as important as drawing is talking about it! Whether counting sheep or piglets in a pen, these mini-maths chats help your child form big number ideas.

Question 1

There are 6 cows in the top field and 4 more in the lower field. How many cows in total?

Build it!

First, place 6 objects on a ten-frame. Add 4 more.

Draw it!

Then, draw 6 dots on the ten-frame. Add 4 more.

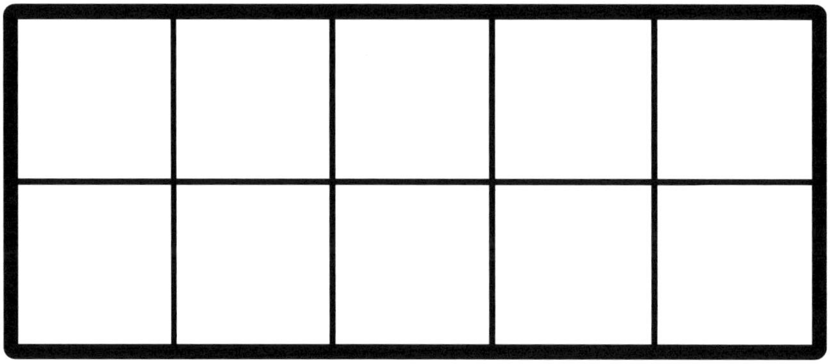

Solve it!

Now, there are ☐ cows altogether.

Question 2

There are 7 cows resting under the trees and 3 more in the barn.
How many cows are there altogether?

Build it!

First, place 7 objects on a ten-frame. Add 3 more.

Draw it!

Then, draw 7 dots on the ten-frame. Add 3 more.

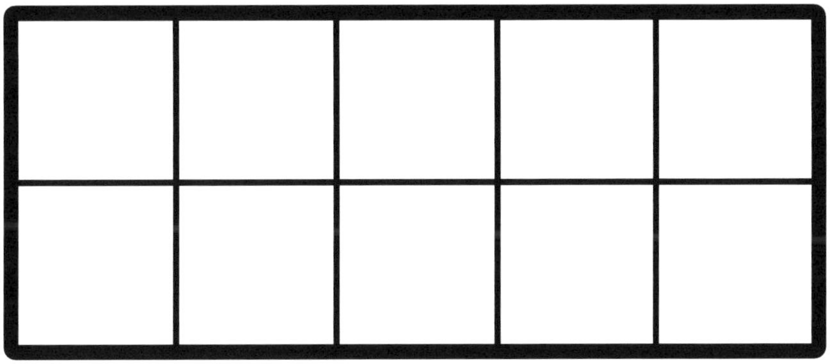

Solve it!

Now, there are ☐ cows in total.

Question 3

There are 12 sheep in the pen.
5 are taken to a field.
How many are left in the pen?

Build it!

First, place 12 objects on two ten-frames. Remove 5.

Draw it!

Then, draw 12 dots on the ten-frames. Cross out 5.

Solve it!

Now, there are ☐ sheep in the pen.

Question 4

There are 12 sheep in the pen.
6 are taken away.
How many are left now?

Build it!

First, place 12 objects on two ten-frames. Remove 6.

Draw it!

Then, draw 12 dots on the ten-frames. Cross out 6.

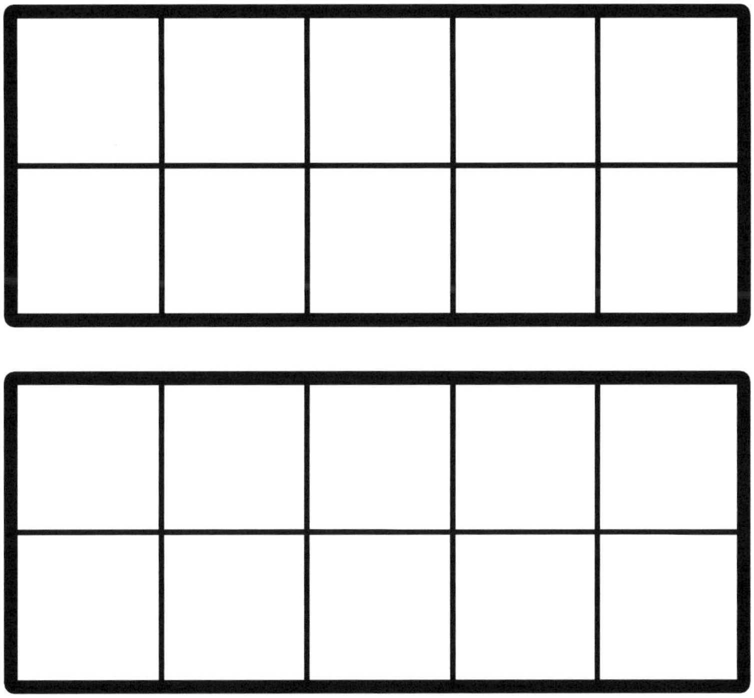

Solve it!

Now, there are ☐ sheep left

Question 5

5 piglets are playing in the mud.
Then 6 more join them.
How many piglets are playing now?

Build it!

First, place 5 objects on a ten-frame. Add 6 more.

Draw it!

Then, draw 5 dots on the first ten-frame. Add 6 more.

Solve it!

Now, ☐ piglets are playing in the mud.

Question 6

11 piglets were playing in the mud.
6 ran to the food trough.
How many are still in the mud?

Build it!

First, place 11 objects on two ten-frames. Remove 6.

Draw it!

Then, draw 11 dots on the ten-frames. Cross out 6.

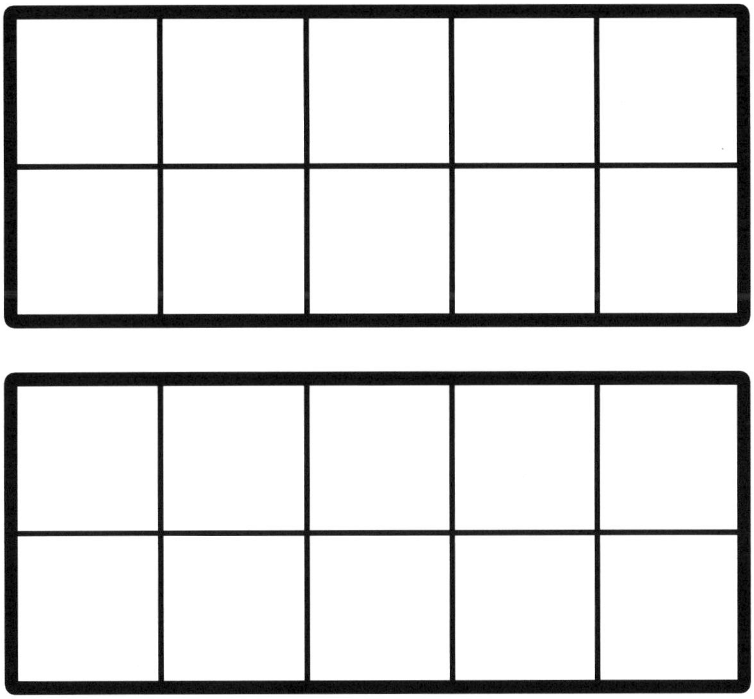

Solve it!

Now, there are ☐ piglets in the mud.

Question 7

There are 8 ducks by the pond.
4 more ducks waddle over.
How many ducks are there now?

Build it!

First, place 8 objects on a ten-frame. Add 4 more.

Draw it!

Then, draw 8 dots on the ten-frame. Add 4 more.

Solve it!

Now, there are ☐ ducks by the pond.

Question 8

12 ducks were at the pond.
4 flew away.
How many ducks remain?

Build it!

First, place 12 objects on two ten-frames. Remove 4.

Draw it!

Then, draw 12 dots on the ten-frames. Cross out 4.

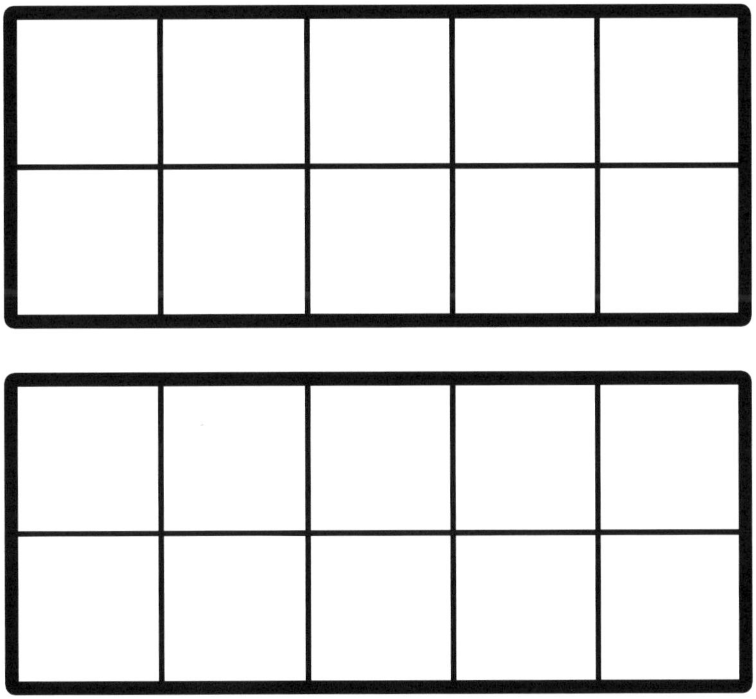

Solve it!

Now, there are ☐ ducks in total.

Question 9

There were 8 frogs in the pond.
5 more joined them, then 3 hopped away.
How many frogs are there now?

Build it!

First, place 8 objects on a ten-frame. Add 5 more, then remove 3.

Draw it!

Then, draw 8 dots on the ten-frame. Add 5 more, then cross out 3.

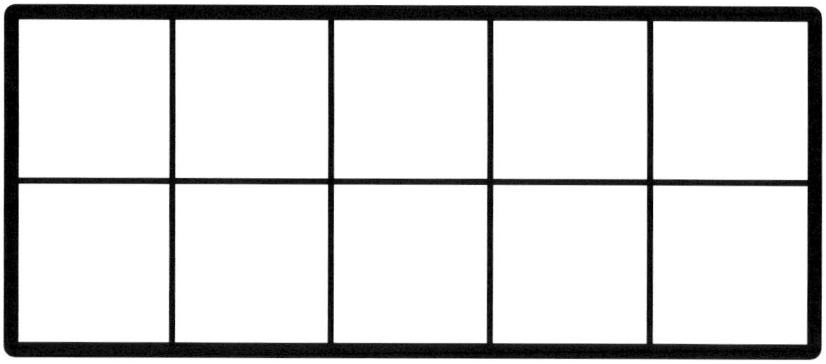

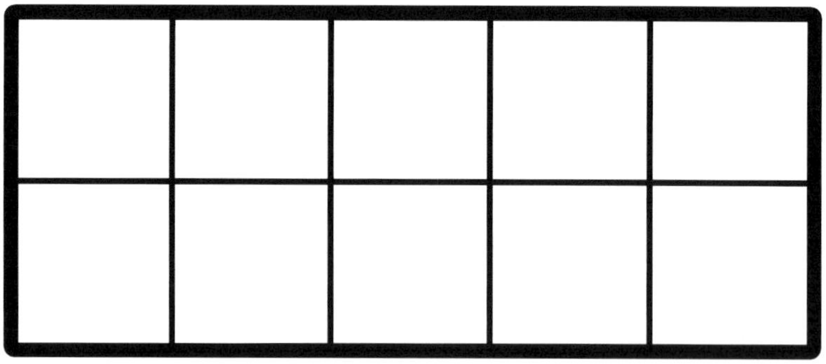

Solve it!

Now, there are ☐ frogs in the pond.

Question 10

There were 8 frogs in the pond.
3 hopped away, then 5 more jumped in.
How many frogs are there now?

Build it!

First, place 8 objects on a ten-frame. Remove 3, then add 5 more.

Draw it!

Then, draw 8 dots on the ten-frame. Cross out 3, then draw 5 more.

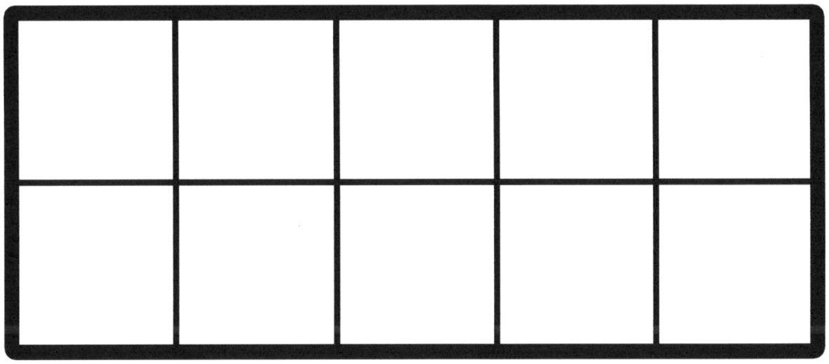

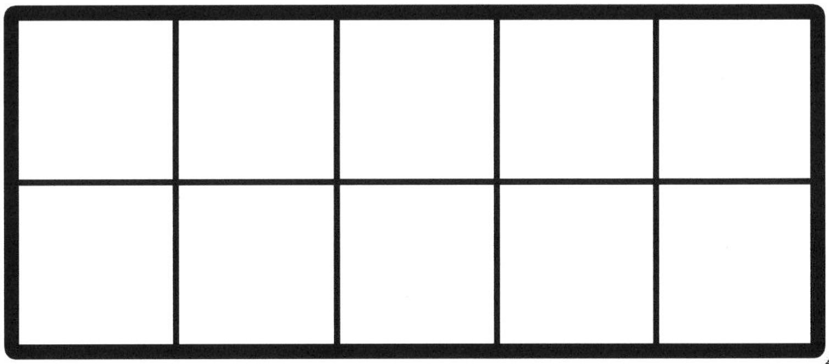

Solve it!

Now, there are ☐ frogs in total.

Write it!

Have ready:

- 2 ten-frames
- objects or counters (20)
- number cards (0-20)
- number word cards (10-20)
- something to write with

Top Tip!

Use phrases such as *tens, ones, place value, worth, number sentence,* and *digits* to help your child describe how numbers fit together.

Write it!

In this step, your child moves from using objects (concrete) and pictures (pictorial) to writing 2-digit numbers and words (**abstract**).

Ten-frames show that every teen number in this section is made from one group of ten and a few extra ones — a big step in understanding place value. They also notice that the position of digits matters — for example, in 14, the '1' means one ten and the '4' means four ones. Matching number cards and words (like 'fourteen') strengthens this understanding. *For example ...*

First, find number cards that match the ten-frames.

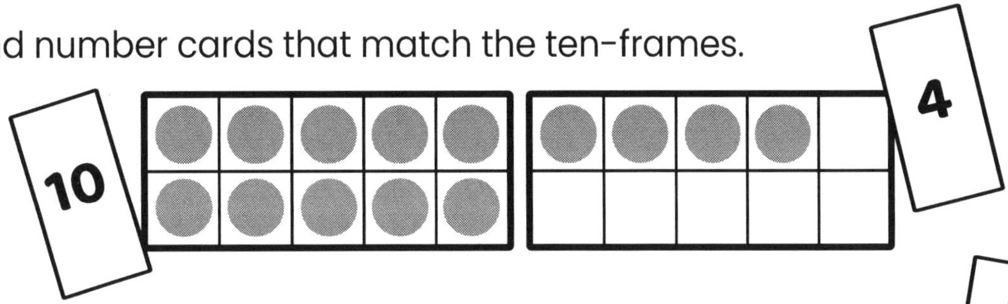

Then, pick up the cards and place 4 on top of the zero in 10.

Now, find the matching number word card.

Each page includes:

- **Write it:** Writing number words
- **Solve it:** Finding missing numbers

And just like every part of this resource, chatting about what's happening or what might happen next helps your child understand how numbers fit together and grow.

Question 1

There are 8 astronauts on a rocket, and 3 more climb aboard.

How many astronauts are on the rocket now?

Build it!

First, place 8 objects on a ten-frame. Add 3 more.

Draw it!

Then, draw 8 dots on the first ten-frame. Add 3 more.

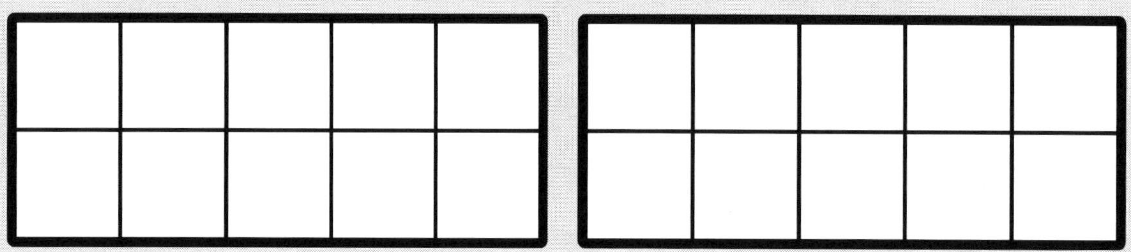

Now, find number cards that match the ten-frames.

Pick up the cards and place 1 on top of the zero in 10.

Write it!

There are ________________ astronauts altogether.

Solve it!

10 and 1, that makes ☐

Question 2

There are 9 stars shining in the sky,
and 3 more start to twinkle.

How many stars can we see altogether?

Build it!

First, place 9 objects on a ten-frame. Add 3 more.

Draw it!

Then, draw 9 dots on the first ten-frame and add 3 more.

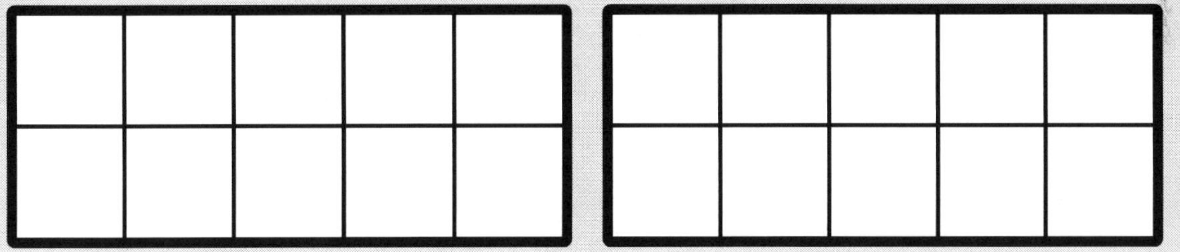

Now, find number cards that match the ten-frames.

Pick up the cards and place 2 on top of the zero in 10.

Write it!

We can see ______________ stars in total.

Solve it!

10 and 2, that makes ☐

Question 3

One train carriage has 9 passengers, and another has 4.

How many passengers are there in total?

Build it!

First, place 9 objects on a ten-frame. Add 4 more.

Draw it!

Then, draw 9 dots on the first ten-frame and draw 3 more.

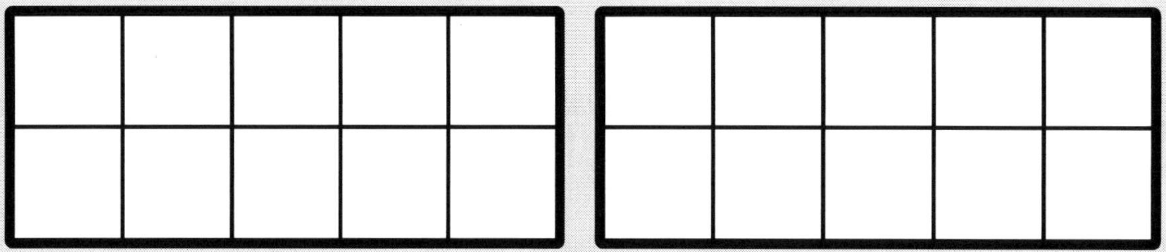

Now, find number cards that match the ten-frames.

Pick up the cards and place 3 on top of the zero in 10.

Write it!

There are ________________ passengers in total.

Solve it!

10 and ☐ , that makes ☐

Question 4

There are 8 cars in one parking bay and 6 in another.

How many cars are there altogether?

Build it!

First, place 8 objects on a ten-frame. Add 6 more.

Draw it!

Then, draw 8 dots on the first ten-frame. Add 6 more.

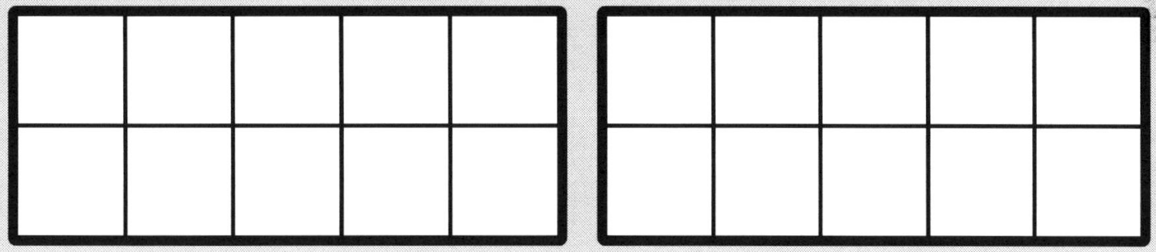

Now, find number cards that match the ten-frames.

Pick up the cards and place 4 on top of the zero in 10.

Write it!

Now, there are ________________ cars altogether.

Solve it!

10 and [   ], that makes [   ]

Question 5

There are 7 rockets on the launch pad, and 8 more land next to them.

How many rockets are there now?

Build it!

First, place 7 objects on a ten-frame. Add 8 more.

Draw it!

Then, draw 7 dots on the first ten-frame and draw 8 more.

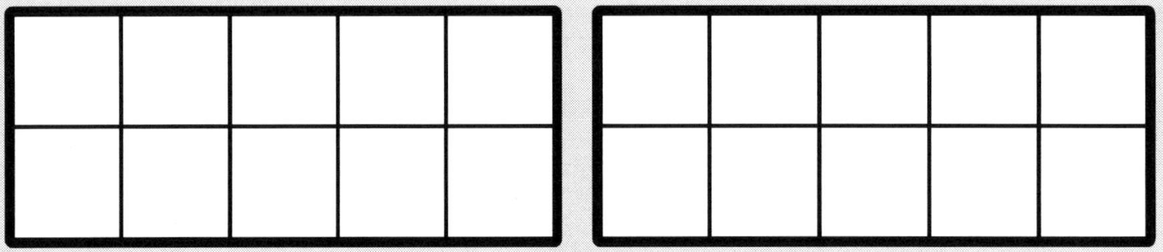

Now, find number cards that match the ten-frames.

Pick up the cards and build the number.

Write it!

There are ______________ rockets in total.

Solve it!

10 and ☐, that makes ☐

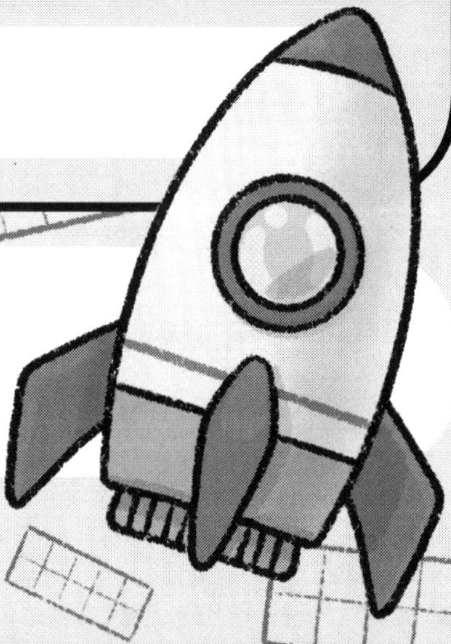

Question 6

A spaceship has 17 aliens on board, and 1 gets off.

How many aliens are still inside?

Build it!

First, place 17 objects on two ten-frames. Remove 1.

Draw it!

Then, draw 17 dots on the ten-frames. Cross out 1.

Now, find number cards that match the ten-frames.

Pick up the cards and build the number.

Write it!

There are ______________ aliens inside the spaceship.

Solve it!

10 and [ ], that makes [ ]

Question 7

19 parcels were loaded onto a delivery van, but 2 fell off.

How many parcels are still on the van?

Build it!

First, place 19 objects on two ten-frames. Remove 2.

Draw it!

Then, draw 19 dots on the ten-frames. Cross out 2.

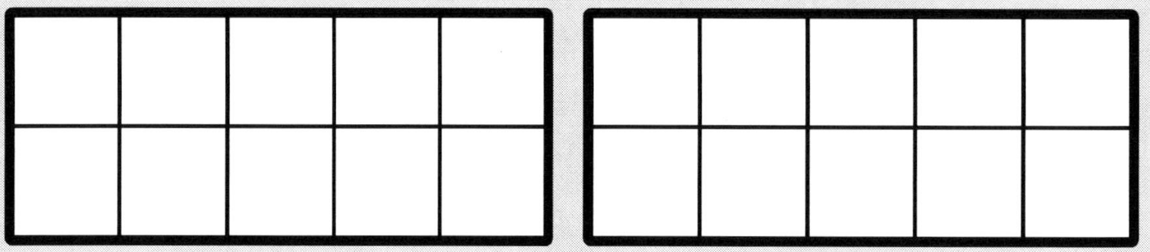

Now, find number cards that match the ten-frames.

Pick up the cards and build the number.

Write it!

Now, _____________ parcels are on the van.

Solve it!

10 and ☐, that makes ☐

Question 8

There are 10 passengers on one minibus and 8 on another.

How many passengers in total?

Build it!

First, place 10 objects on a ten-frame. Add 8 more.

Draw it!

Then, draw 10 dots on the first ten-frame and draw 8 more.

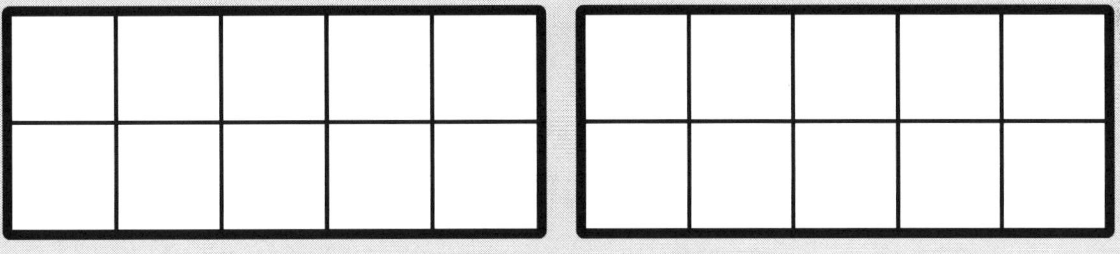

Now, find number cards that match the ten-frames.

Pick up the cards and build the number.

Write it!

There are ______________ passengers altogether.

Solve it!

10 and ☐, that makes ☐

Question 9

The astronauts loaded 20 moon rocks into a rocket, but 1 rolled out.

How many moon rocks are still in the rocket?

Build it!

First, place 20 objects on two ten-frames. Remove 1.

Draw it!

Then, draw 20 dots. Cross out 1.

Now, find number cards that match the ten-frames.

Pick up the cards and build the number.

Write it!

There are _____________moon rocks in the rocket.

Solve it!

10 and ______, that makes ______

Question 10

There were 15 aliens on a planet, but 5 flew away on a spaceship.

How many aliens are left?

Build it!

First, place 15 objects on two ten-frames. Remove 5.

Draw it!

Then, draw 15 dots. Cross out 5.

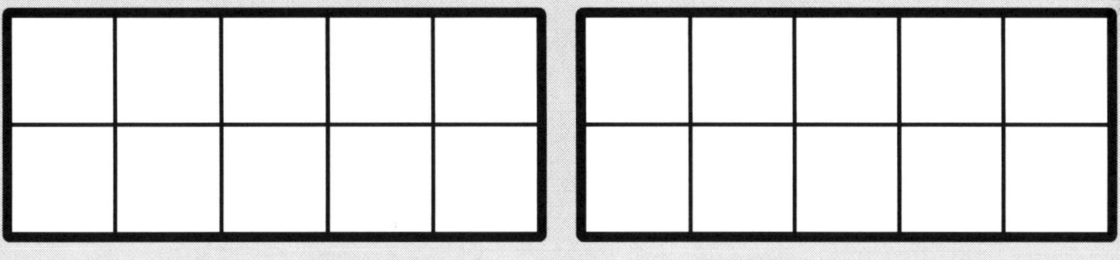

Now, find number cards that match the ten-frames.

Pick up the cards and build the number.

Write it!

There are _____________aliens on the planet.

Solve it!

10 and ☐ , that makes ☐

Group it! Share it!

Have ready:

- 6 ten-frames
- objects or counters (30)
- something to write with

Top Tip!
Use words like *groups of, lots of, multiply, divide, altogether* and *shared equally* while your child explains what's happening.

Group it! Share it!

In this step, your child uses what they already know about adding and subtracting to explore multiplication and division. They'll group and share objects using real ten-frames and drawings before moving to number sentences.

These activities continue the CPA approach — moving from real objects (concrete) to drawings (pictorial) and then to numbers and symbols (abstract).

And of course, chatting about what they notice supports spotting patterns and making connections between operations – key skills for young mathematicians!

Using Ten-Frames to Support Multiplication

Ten-frames help children see multiplication as **repeated addition** by letting them count equal groups in a clear, visual way.

Example:

Repeated addition
The baker places 3 cupcakes on each tray. She uses 5 trays. How many cupcakes does she have in total?
The child places 3 objects in each ten-frame: 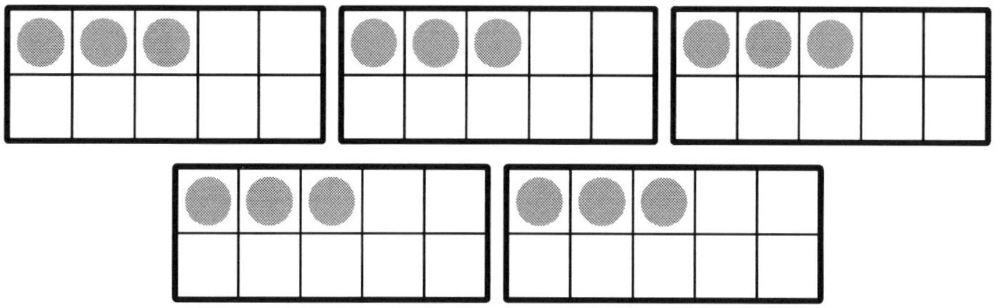
Answer: 3 × 5 = 15 cupcakes.
Why it matters: Children see that 3 added five times (3 + 3 + 3 + 3 + 3) equals 15, just like 3 × 5.

Whether your child is arranging cookies in boxes or counting cupcakes in rows, these baking-themed activities make learning multiplication fun while strengthening their number sense.

Question 1

The baker puts two cookies in each box.
She packs five boxes.

How many cookies does the baker have in total?

Build it!

First, have ready ☐ ten-frames. Place ☐ objects on each.

Draw it!

Then, draw ☐ dots in each of the ten-frames.

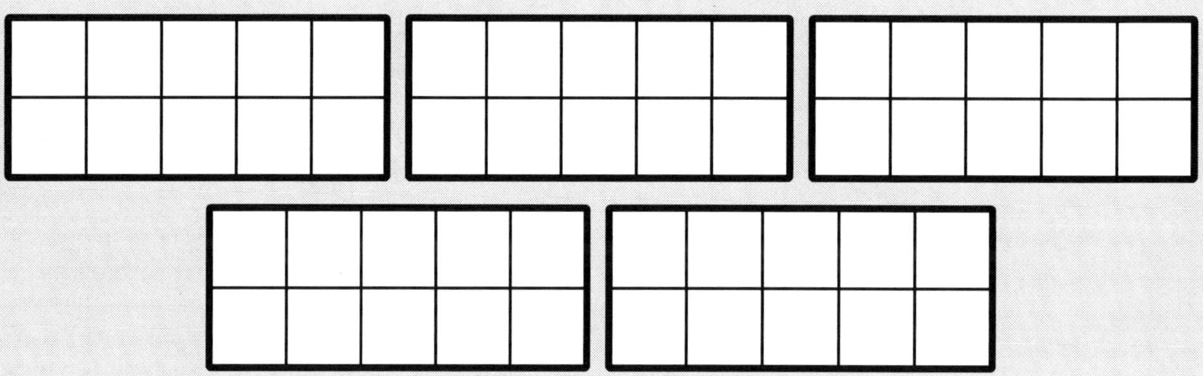

Write it!

Now, the baker has _________cookies.

Solve it!

2 times 5 equals ☐

Question 2

The baker puts five cookies in each box.
She packs two boxes.

How many cookies does the baker have altogether?

Build it!

First, have ready ☐ ten-frames. Place ☐ objects on each.

Draw it!

Then, draw ☐ dots in each of the ten-frames.

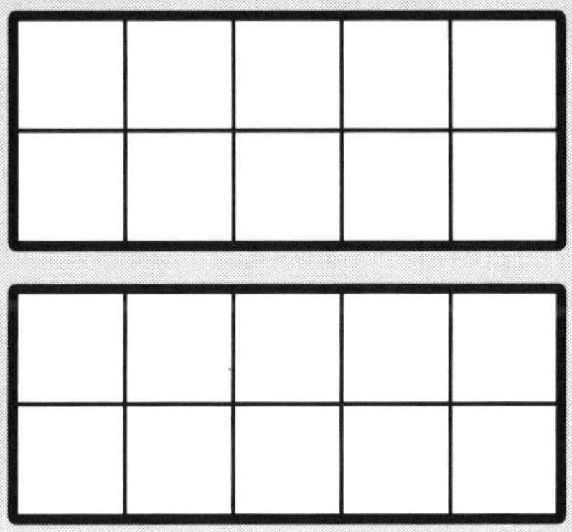

Write it!

Now, the baker has ________cookies.

Solve it!

5 multiplied by 2, that makes ☐

Question 3

The baker lines up three rows of cupcakes with four in each row.

How many cupcakes does the baker have in total?

Build it!

First, have ready ☐ ten-frames. Place ☐ objects on each row.

Draw it!

Then, draw ☐ dots in each row of the ten-frames.

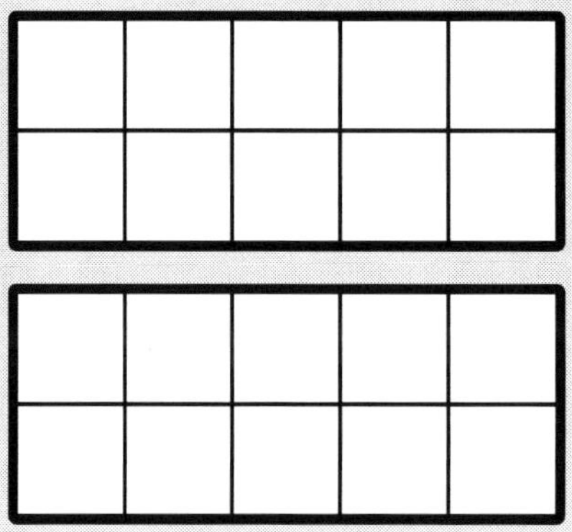

Write it!

Now, the baker has ________cupcakes.

Solve it!

4 times 3 equals ☐

Question 4

The baker lines up four rows of cupcakes with three in each row.

How many cupcakes does the baker have altogether?

Build it!

First, have ready ☐ ten-frames. Place ☐ objects on each.

Draw it!

Then, draw ☐ dots in each of the ten-frames.

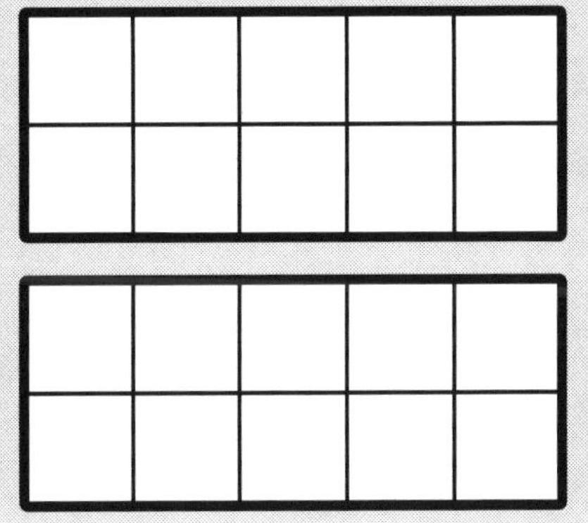

Write it!

Now, the baker has ________cupcakes.

Solve it!

3 x 4 = ☐

Question 5

The baker uses six trays and puts five muffins on each tray.

How many muffins does the baker have in total?

Build it!

First, have ready ☐ ten-frames. Place ☐ objects on each.

Draw it!

Then, draw ☐ dots in each of the ten-frames.

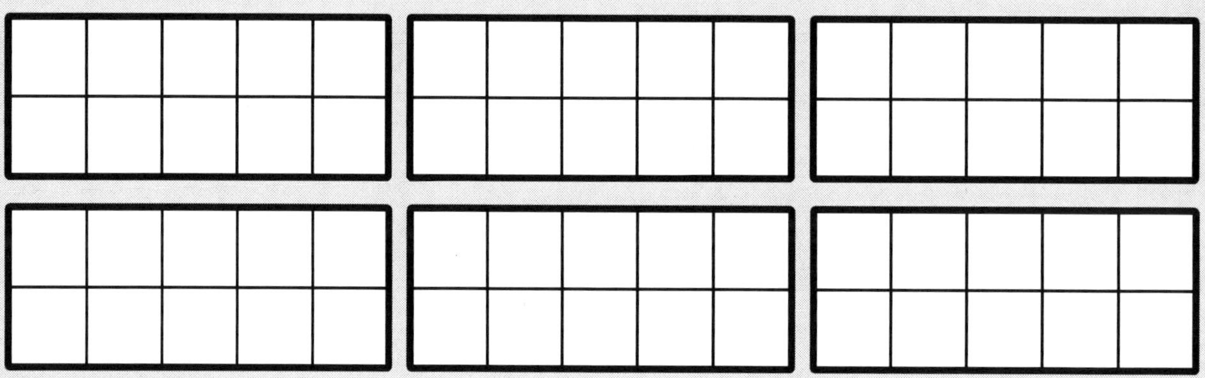

Write it!

Now, the baker has ________muffins.

Solve it!

5 multiplied by 6 equals ☐

Question 6

The baker uses five trays and puts six muffins on each tray.

How many muffins does the baker have altogether?

Build it!

First, have ready ☐ ten-frames. Place ☐ objects on each.

Draw it!

Then, draw ☐ dots in each of the ten-frames.

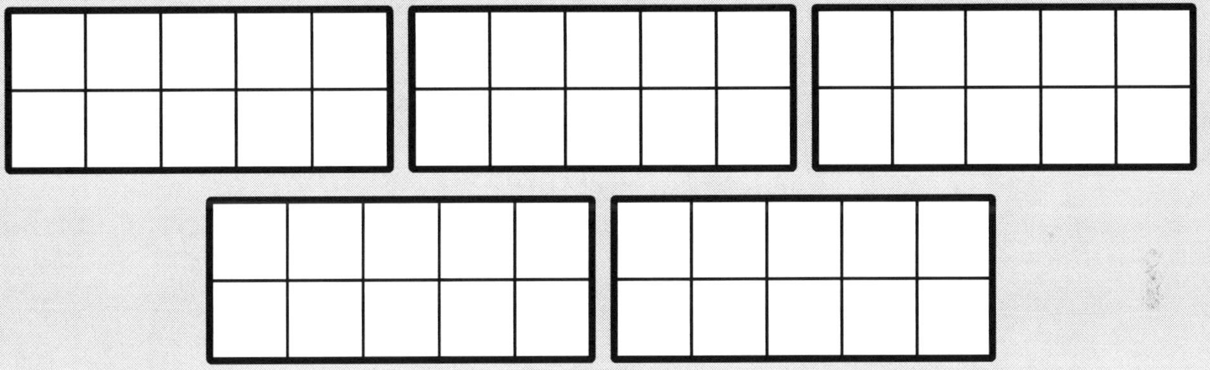

Write it!

Now, the baker has ________muffins.

Solve it!

6 x 5 = ☐

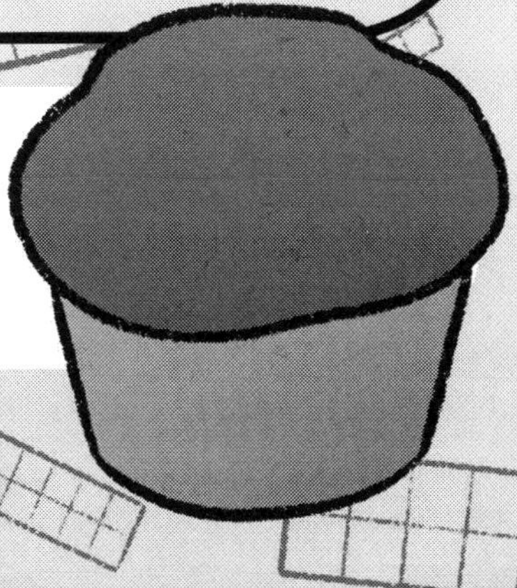

Question 7

The baker fills three trays
with five buns on each.

How many buns does the baker have in total?

Build it!

First, have ready ☐ ten-frames. Place ☐ objects on each.

Draw it!

Then, draw ☐ dots in each of the ten-frames.

Write it!

Now, the baker has ________buns.

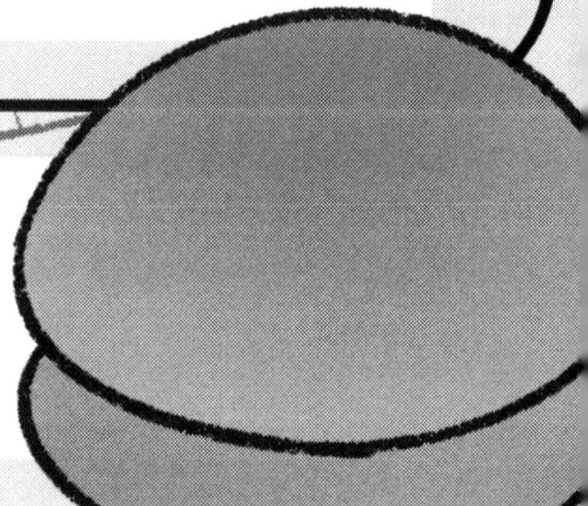

Solve it!

5 x 3 = ☐

Question 8

The baker fills five trays with three buns on each.

How many buns does the baker have altogether?

Build it!

First, have ready [ ] ten-frames. Place [ ] objects on each.

Draw it!

Then, draw [ ] dots in each of the ten-frames.

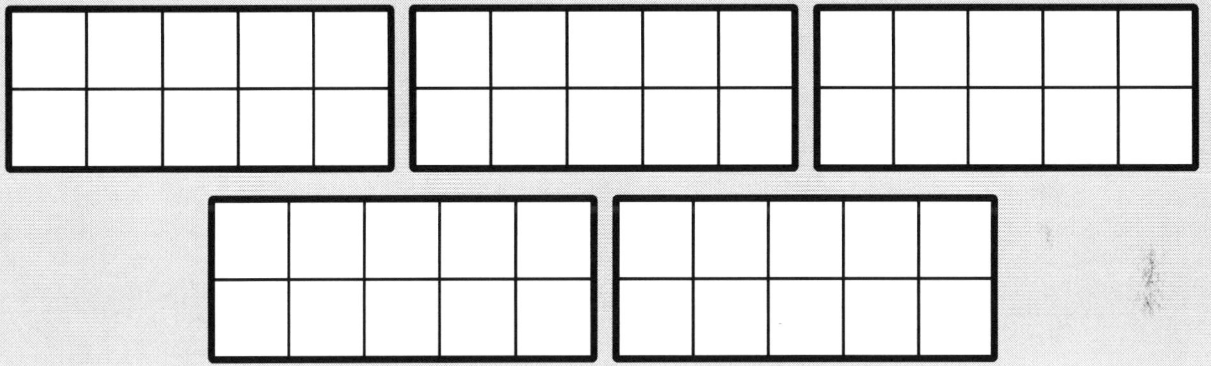

Write it!

Now, the baker has _________buns in total.

Solve it!

3 times 5, that makes [ ]

Question 9

The baker fills four trays with six bagels each.

How many bagels does the baker bake in total?

Build it!

First, have ready [ ] ten-frames. Place [ ] objects on each.

Draw it!

Then, draw [ ] dots in each of the ten-frames.

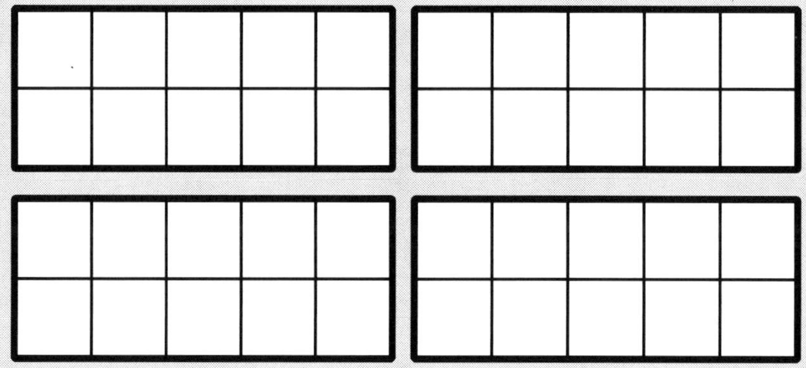

Write it!

Now, the baker has ________bagels altogether.

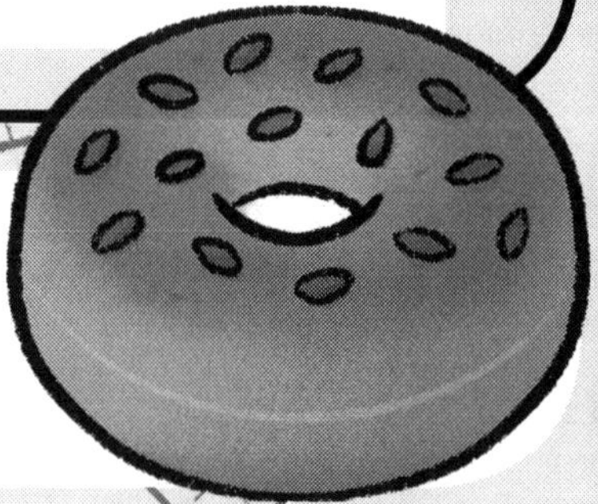

Solve it!

6 x 4 = [ ]

Question 10

The baker fills six trays with four bagels each.

How many bagels does the baker have altogether?

Build it!

First, have ready ☐ ten-frames. Place ☐ objects on each.

Draw it!

Then, draw ☐ dots in each of the ten-frames.

Write it!

Now, the baker has ________bagels in total.

Solve it!

4 multiplied by 6, that makes ☐

Using Ten-Frames to Support Division

For division, ten-frames make it easy to share or group items equally so that children begin to see that division can mean **sharing equally** or how many **equal groups** of something we can make.

Example:

Sharing equally
The baker has 10 cookies and wants to put them equally on 2 plates. How many cookies will go on each plate?
Child shares 10 objects one by one between 2 ten-frames: 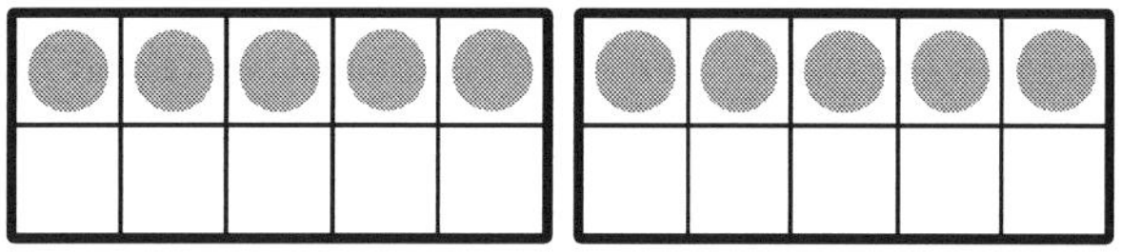
Answer: 10 ÷ 2 = 5 cookies per plate.
Why it matters: This helps children understand how to divide fairly.

Equal groups
The baker has 10 cookies. He puts 2 cookies in each bag. How many bags in total?
Child draws 10 dots in the ten-frame and circles each group of 2: 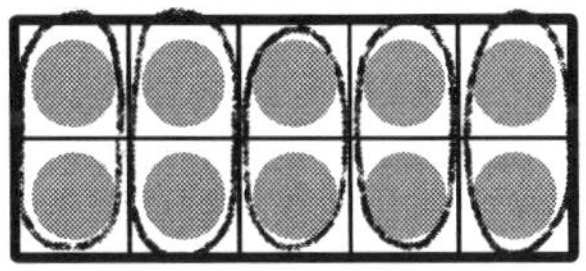
Answer: 10 ÷ 2 = 5 bags altogether.
Why it matters: This helps children understand division can also mean *how many groups* of something we can make.

Whether your child is sharing cookies into bags or putting 4 cupcakes on each plate, these baking-themed activities help them explore division and build strong number sense.

Question 1

The baker made ten cupcakes
and wants to share them equally between two plates.

How many cupcakes go on each plate?

Build it!

First, use 2 ten-frames. Share 10 objects one by one between them.

Draw it!

Then, draw 10 dots, sharing them evenly across the 2 ten-frames.

Write it!

Now, there are _________ cupcakes on each plate.

Solve it!

10 shared by 2, that makes ☐

Question 2

The bakery has ten cupcakes.
Each cookie box holds two cupcakes.

How many boxes can they fill?

Build it!

First, use a ten-frame to place 10 objects. Group into sets of 2.

Draw it!

Then, draw 10 dots in the ten-frame and circle each group of 2.

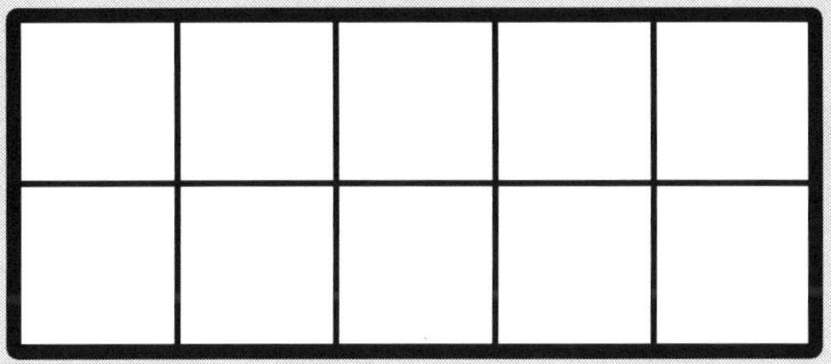

Write it!

Now, there are ________ cupcakes in each box.

Solve it!

10 divided by 2, equals ☐

Question 3

There are ten doughnuts.
They are shared equally between five children.

How many does each child get?

Build it!

First, use 5 ten-frames. Share ____ objects one by one between them.

Draw it!

Then, draw ____ dots, sharing them evenly across the 5 ten-frames.

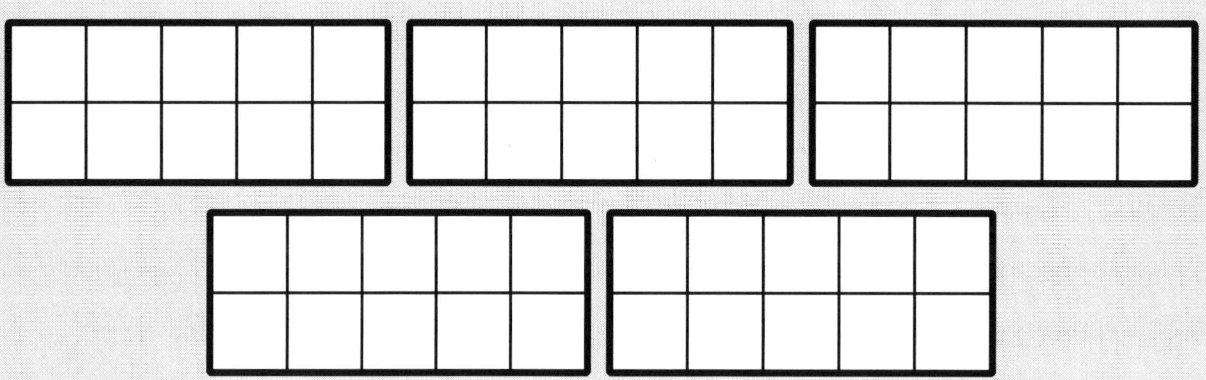

Write it!

Now, each child has _______ doughnuts.

Solve it!

$10 \div 5 =$ ____

Question 4

A baker packs ten doughnuts into boxes with five doughnuts in each.

How many boxes does she fill?

Build it!

First, use a ten-frame to place 10 objects. Group them into sets of 5.

Draw it!

Then, draw ☐ dots and circle each group of 5.

Write it!

Now, there are ________ doughnuts in each box.

Solve it!

10 divided by 5, equals ☐

Question 5

The baker made twenty buns
and wants to share them equally between four trays.

How many buns go on each tray?

Build it!

First, use 4 ten-frames. Share ☐ objects one by one across them.

Draw it!

Then, draw ☐ dots, sharing them evenly across the 4 ten-frames.

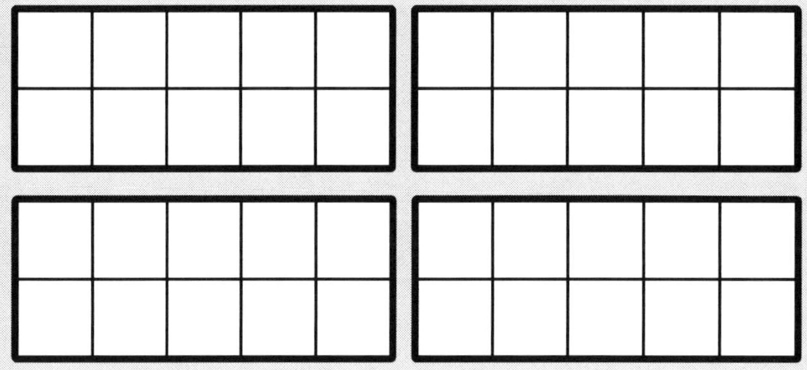

Write it!

Now, there are ________ buns on each tray.

Solve it!

20 shared by 4, that makes ☐

Question 6

The bakery has twenty buns.
Each tray holds four buns.

How many trays can they fill?

Build it!

First, use ten-frames to place 20 objects. Group into sets of ☐

Draw it!

Then, draw 20 dots in the ten-frames and circle each group of ☐

Write it!

Now, each tray has ________ buns.

Solve it!

20 ÷ 4 = ☐

Question 7

The baker has twelve muffins
and wants to share them equally between four trays.

How many muffins go on each tray?

Build it!

First, use 4 ten-frames. Share objects one by one across them.

Draw it!

Then, draw dots and share them evenly across 4 ten-frames.

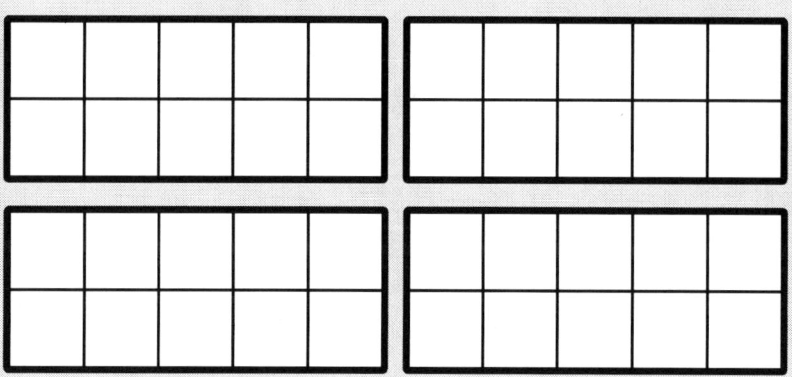

Write it!

Now, there are ________ muffins on each tray.

Solve it!

$12 \div 4 =$

Question 8

The baker has twelve muffins.
She puts four muffins in each box.

How many boxes does she fill?

Build it!

First, use ten-frames to place ☐ objects. Group them into sets of 4.

Draw it!

Then, draw 12 dots and circle each group of ☐

Write it!

Now, there are ________ boxes in total.

Solve it!

12 divided by 4 equals ☐

Question 9

The baker has fifteen cookies to share equally between three trays.

How many cookies go on each tray?

Build it!

First, use 3 ten-frames. Share ☐ objects one by one across them.

Draw it!

Then, draw ☐ dots, sharing them evenly across the 3 ten-frames.

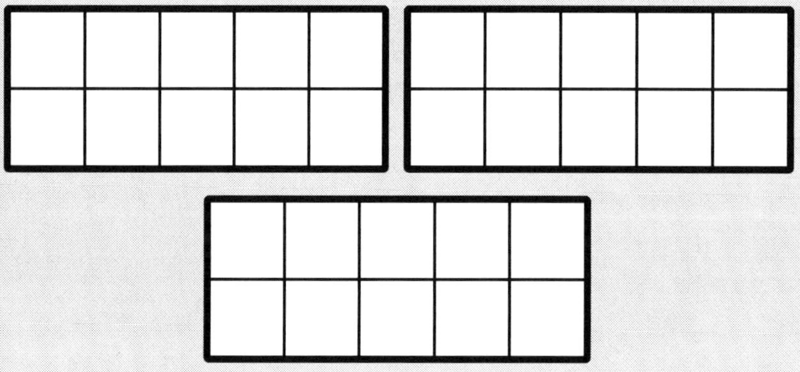

Write it!

Now, there are ________ cookies on each tray.

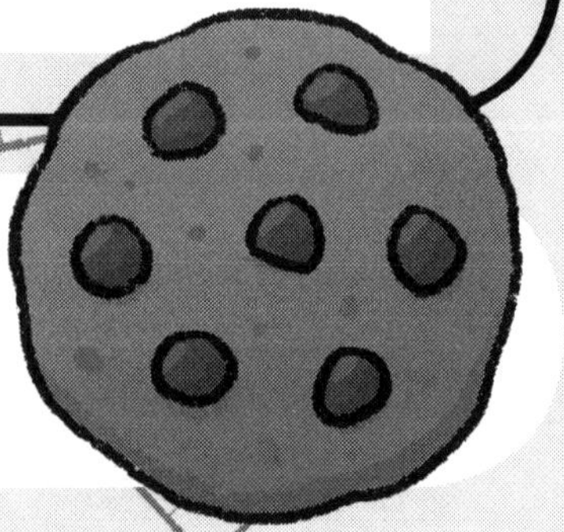

Solve it!

15 shared by 3, that makes ☐

Question 10

The baker has fifteen cookies
and puts three cookies in each bag.

How many bags does she make?

Build it!

First, use ten-frames to place [ ] objects. Group them into sets of [ ]

Draw it!

Then, draw 15 dots and circle each group of [ ]

Write it!

Now, there are ________ cookies on each tray.

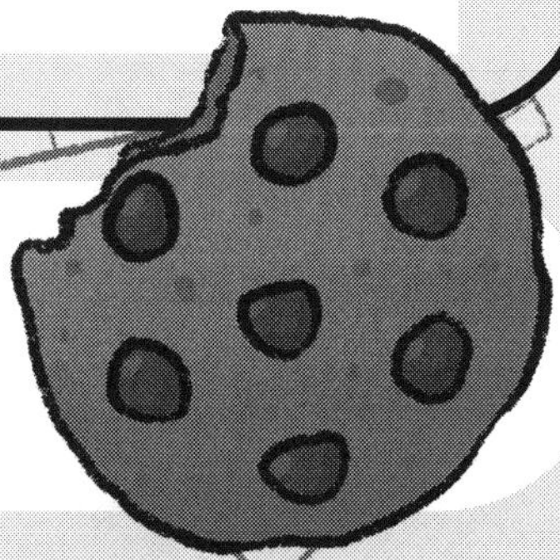

Solve it!

$15 \div 3 =$ [ ]

Apply it!

Have ready:

- 4 ten-frames
- objects or counters (20)
- something to write with

Top Tip!
Use key operation words like *add, subtract, multiply, divide, share equally, groups of, total* and *altogether* when talking about number problems together.

Apply it!

In this final step, your child brings everything together — addition, subtraction, multiplication, and division — to solve everyday problems. They'll write numbers, find missing parts, and use and explain their thinking, building flexible and confident problem-solving skills.

These activities connect hands-on learning (like using objects and pictures) with written maths and help build number sense.
Each challenge helps your child apply ideas in new ways — a key to flexible thinking.

Talking about what you see helps your child reason and notice number patterns.
For example: If they have 12 counters and you take 4 away, ask:

These small conversations help your child make sense of the numbers, and grow their confidence too!

Question 1

Jen saw twelve stars.
Then she counted five more in her telescope.

How many stars did she see in total?

Build it!

First, place 12 objects on two ten-frames. Add 5 more.

Draw it!

Then, draw 12 dots on the first ten-frames. Add 5 more.

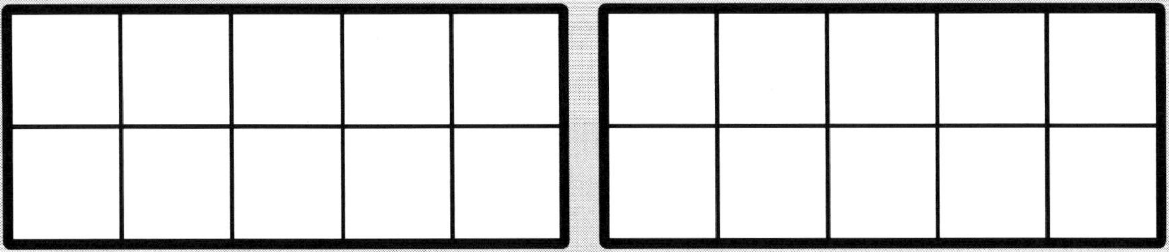

Write it!

Now, Jen can see ___________ stars altogether.

Solve it!

12 add 5, that makes ☐

Question 2

Jayden saw thirteen stars.
Then he spotted four more.

How many stars altogether?

Build it!

First, use objects for the first number.
Then add more counters to make the next one."

Draw it!

Then, draw dots for both parts.

Write it!

Now, there are ____________ stars in total.

Solve it!

13 plus 4, equals ☐

Question 3

Eighteen boats were in the harbour. Six sailed away. How many boats are in the harbour now?

Build it!

First, place 18 objects on two ten-frames. Remove 6.

Draw it!

Then, draw 18 dots on the ten-frames. Cross out 6.

Write it!

Now, there are ____________ boats in the harbour.

Solve it!

18 take away 6, that makes ☐

Question 4

Eighteen boats were in the harbour. Four sailed away.
How many boats are in the harbour now?

Build it!

First, start with 18 objects. Take some away.

Draw it!

Then, draw 18 dots on the ten-frames. Cross out 4.

Write it!

Now, there are _____________boats in the harbour.

Solve it!

18 minus 4, equals ☐

Question 5

There are four rockets.
Each rocket carries three astronauts.

How many astronauts in total?

Build it!

First, have ready ☐ ten-frames. Place ☐ objects on each.

Draw it!

Then, draw ☐ dots in each of the ten-frames.

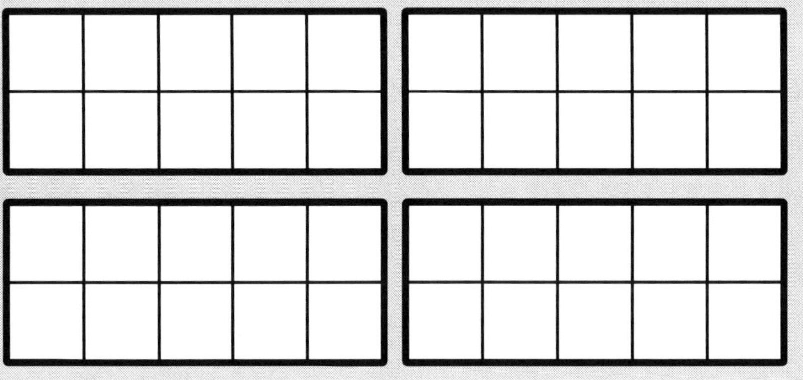

Write it!

Now, there are ________ astronauts altogether.

Solve it!

3 times 4, that makes ☐

Question 6

There are two rockets.
Each has six astronauts on board.

How many astronauts are there?

Build it!

First, have ready ____ ten-frames. Place ____ objects on each.

Draw it!

Then, draw ____ dots in each of the ten-frames.

Write it!

Now, there are _________ astronauts altogether.

Solve it!

6 multiplied by 2 equals ____

Question 7

Sixteen wheels need to be shared equally between four toy lorries.

How many wheels does each lorry get?

Build it!

First, have ready 4 ten-frames and place ☐ objects one by one on each.

Draw it!

Then, draw ☐ dots in each of the ten-frames.

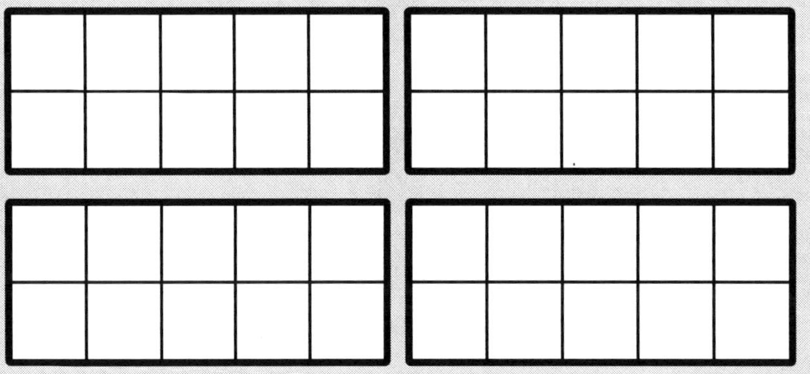

Write it!

Now, there are _________ wheels on each lorry.

Solve it!

16 shared by 4, that makes ☐

Question 8

There are sixteen wheels.
Each lorry uses four wheels.

How many lorries can we make?

Build it!

First, place 16 objects on ten-frames. Group into sets of 4.

Draw it!

Then, draw 16 dots in the ten-frame and circle each group of 4.

Write it!

Now, there are ________ lorries in total.

Solve it!

16 divided by 4 equals ☐

Question 9

Zac saw nine stars, then seven more appeared. Later, three faded away.

How many stars does he see now?

Build it!

First, place 9 objects on a ten-frame. Add 7 more. Next, remove 3.

Draw it!

Then, draw 9 dots on the first ten-frame. Add 7 more. Cross out 3.

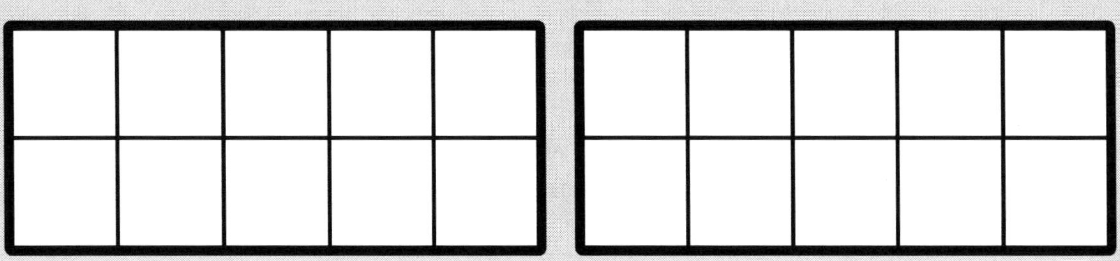

Write it!

Now, Zac can see ____________ stars.

Solve it!

9 add 7, take away 3 ... that makes ☐

Question 10

Rebecca saw ten stars, then five more.
Later, two disappeared.

How many are left?

Build it!

First, place 10 objects on a ten-frame. Add 5 more. Remove 2.

Draw it!

Then, draw 10 dots on the first ten-frame. Add 5 more. Cross out 2.

Write it!

Now, there are ____________ stars left.

Solve it!

10 add 5, subtract 2 ... equals ☐

Question 11

Ravi made six cupcakes in the morning, the same number in the afternoon, and two more in the evening.

How many cupcakes were made in total?

Build it!

First, place 6 objects on a ten-frame. Add the same amount, plus 2 more.

Draw it!

Then, draw 6 dots on the first ten-frame.
Draw the same amount and add 2 more.

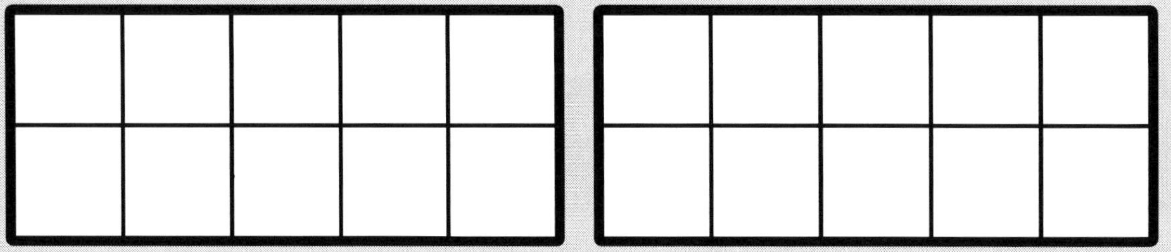

Write it!

Now, there are ____________ cupcakes in total.

Solve it!

Double ☐ is ☐, add 2 more ... that makes ☐

Question 12

Hattie made six cupcakes in the morning
and the same number in the afternoon.
Then two cupcakes were eaten.
How many cupcakes are left?

Build it!

First, place 6 objects on a ten-frame. Add the same amount and remove 2.

Draw it!

Then, draw 6 dots on the first ten-frame.
Draw the same amount and cross out 2.

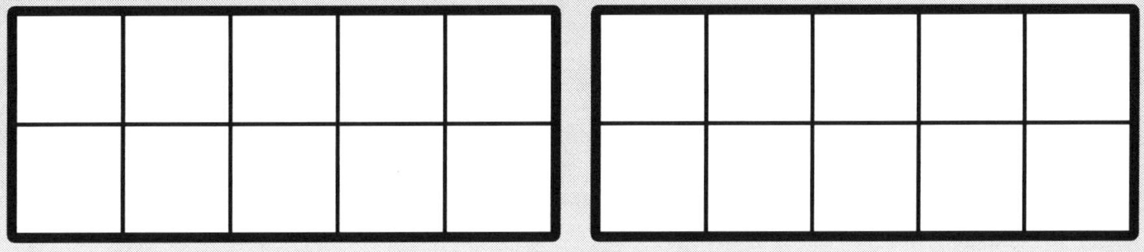

Write it!

Now, there are ____________ cupcakes.

Solve it!

Double ☐ is ☐, subtract 2 ... equals ☐

Question 13

There are seven pigs in the pen,
three pigs in the field, and six pigs by the trough.

How many pigs are there altogether?

Build it!

First, place 7 objects on a ten-frame. Add 3 and 6 more.

Draw it!

Then, draw 7 dots on the first ten-frame. Draw 3 and add 6 more.

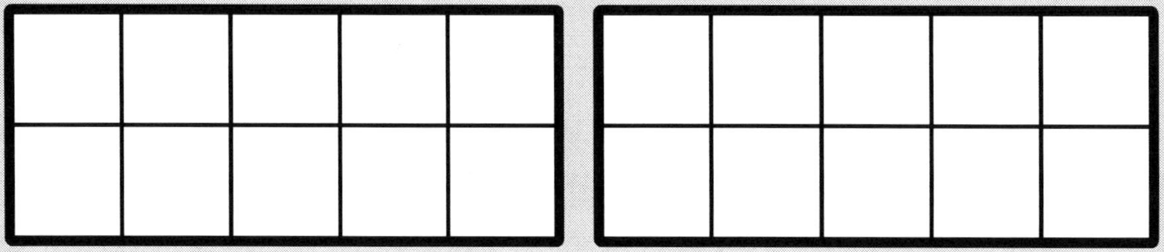

Write it!

Now, there are ____________ pigs altogether.

Solve it!

7 and [ ], plus 6 ... that makes [ ]

Question 14

There are seven pigs in the pen and three pigs in the field. Then six pigs go inside the barn.

How many pigs are left outside?

Build it!

First, place 7 objects on a ten-frame. Add 3 more, then remove 6.

Draw it!

Then, draw 7 dots on the first ten-frame. Draw 3 more, then cross out 6.

Write it!

Now, ____________ pigs are left outside.

Solve it!

☐ and 3, minus 6 ... that makes ☐

Question 15

Marco baked four buns on one tray, the same number on another tray, and the same number again on a third tray.

How many buns were baked altogether?

Build it!

First, place 4 objects on a ten-frame.
Add the same amount on the other ten-frames.

Draw it!

Then, draw ☐ dots on the first ten-frame.
Draw the same amount on the other ten-frames.

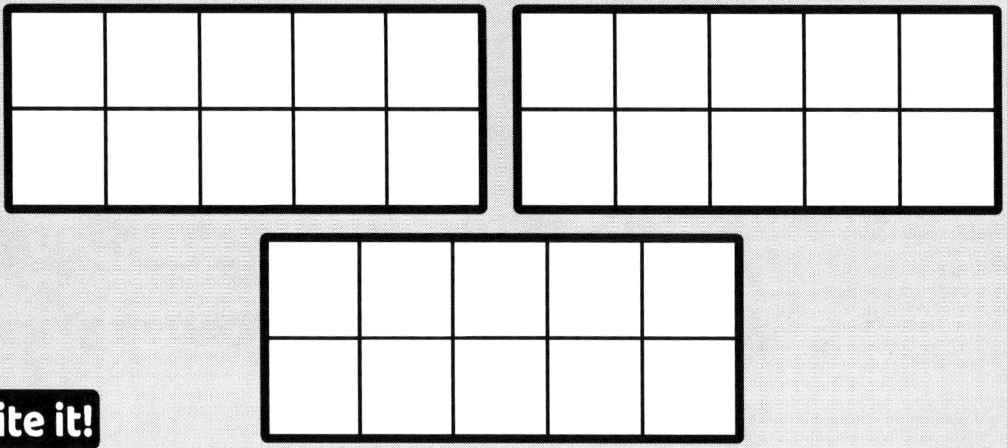

Write it!

Now, there are ____________ buns in total.

Solve it!

4 plus ☐ plus ☐ ...that makes ☐

Question 16

Amira baked four buns on one tray and the same number on another tray. Then the same number were eaten by friends. How many buns are left?

Build it!

First, place 4 objects on a ten-frame.
Add the same amount on the other ten-frame.
Remove the same number.

Draw it!

Then, draw 4 dots on the first ten-frame and the same number on the second ten-frame. Cross out ☐

Write it!

Now, ____________ buns are left.

Solve it!

4 add ☐ subtract ☐ ...equals ☐

Question 17

There are eight ducks on the pond,
two ducks in the yard, and five ducks in the barn.

How many ducks are there altogether?

Build it!

First, place 8 objects on a ten-frame. Add 2 and 5 more.

Draw it!

Then, draw 8 dots on the first ten-frame. Draw 2 and add 5 more.

Write it!

Now, there are ____________ ducks in total.

Solve it!

8 and 2, add ☐ more ... that makes ☐

Question 18

There are eight ducks on the pond
and two ducks in the yard. Then five ducks fly away.

How many ducks are left?

Build it!

First, place 8 objects on a ten-frame. Add 2 more, then remove 5.

Draw it!

Then, draw 8 dots on the first ten-frame. Draw 2 more, then cross out 5.

Write it!

Now, there are ____________ ducks on the pond.

Solve it!

8 plus ☐ minus ☐ ...equals ☐

Question 19

Luca baked five loaves in the morning and the same amount in the afternoon.

How many loaves were baked altogether?

Build it!

First, place 5 objects on a ten-frame and add the same amount.

Draw it!

Then, draw 5 dots on the ten-frame and add the same number.

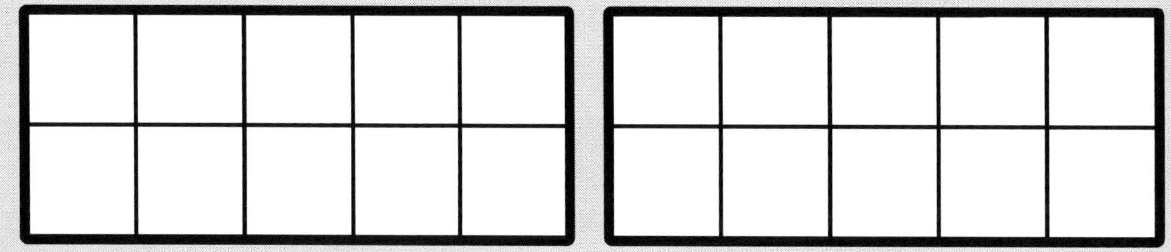

Write it!

Now, there are ____________ loaves in total.

Solve it!

Double [ ] equals [ ]

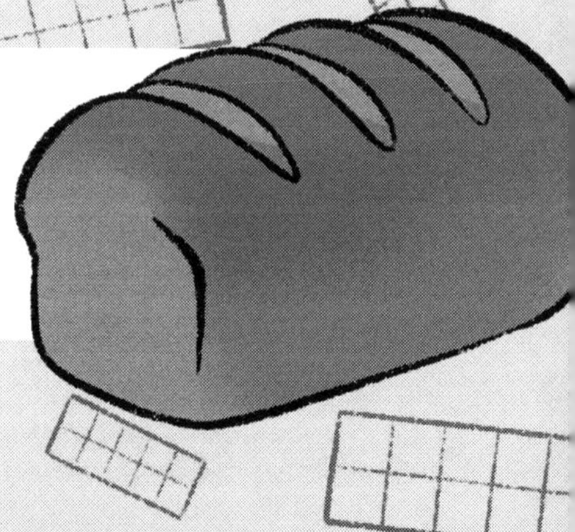

Question 20

Sofia made ten loaves in the morning.
Half were sold in the afternoon.

How many loaves are there now?

Build it!

First, place ☐ objects on a ten-frame. Remove half.

Draw it!

Then, draw ☐ dots on the ten-frame. Cross out half.

Write it!

Now, there are ____________ loaves for sale.

Solve it!

Half of ☐ equals ☐

See it! Say it!

Have ready:

- ten-frame flashcards (0-10)
- number cards (0-19)
- number word cards (ten-twenty)

Top Tips

- It's okay if your child is unsure at first — quick looking takes practice.
- Encourage your child to explain how they saw the number, not just say the answer.
- If your child starts counting, gently cover the dots and ask what they noticed first.
- Use everyday moments (stairs, snacks, toys) to ask "How many can you see?"
- Praise effort and thinking, not speed — confidence comes before quickness.

See it! Say it!

This section builds subitising skills: the ability to see how many without counting.

Ten-frames help with this by showing numbers in clear, organised patterns. When children see dots or counters in certain places, they begin to know "how many" just by looking.

Example: Seeing 3 counters and knowing it's 3.

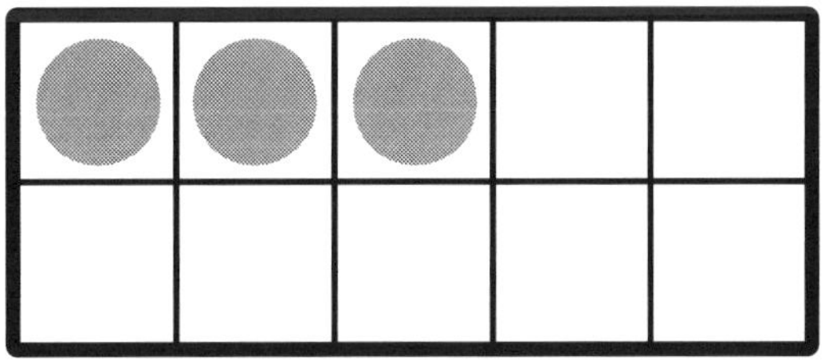

Or noticing patterns — for example, spotting 5 on the top row and realising it's one fewer than 6.

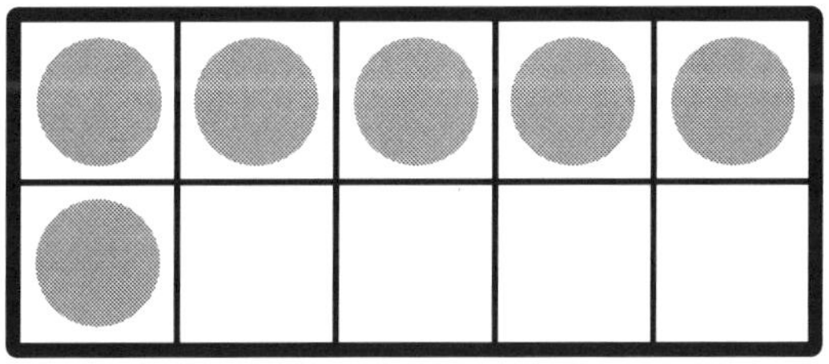

These activities are more than just quick counting — they help your child:

- Recognise numbers quickly by sight
- Spot patterns and relationships between and within numbers
- Build strong number sense

The following games grow with your child — from spotting numbers at a glance to thinking about how numbers fit together. And of course, talking together will help your child describe what they see, use mathematical language, and make connections between numbers — all key parts of confident early maths learning!

The Big Challenge!

Have ready:

- Flashcards (0-10)

How to play:

1. Shuffle the cards and turn one over.

2. Ask: "How many can you see without counting?"

Example: 7

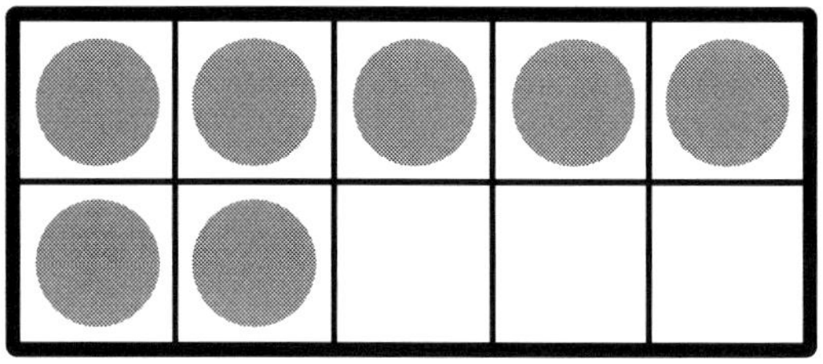

3. Keep going until all cards are turned.

This activity helps your child to recognise small numbers quickly without counting each one.

The Bigger Challenge!

Have ready:

- Flashcards (0-10)
- Number cards (10-19)

1. Place the flashcard showing 10 dots onto a flat surface.
2. Shuffle the remaining flashcards and place face down in a pile.

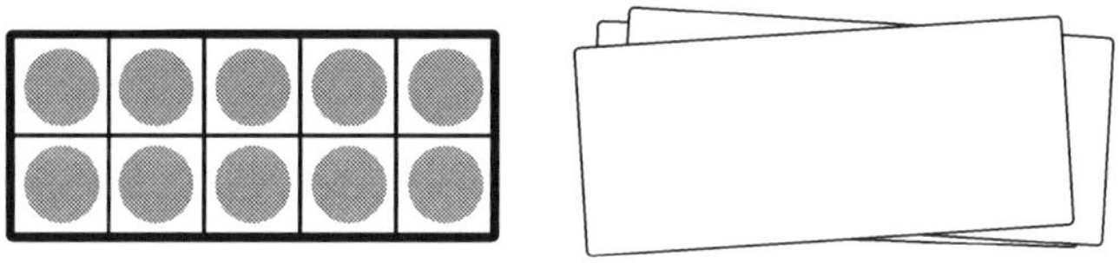

3. Turn one over and ask: *'How many can you see altogether without counting?'*

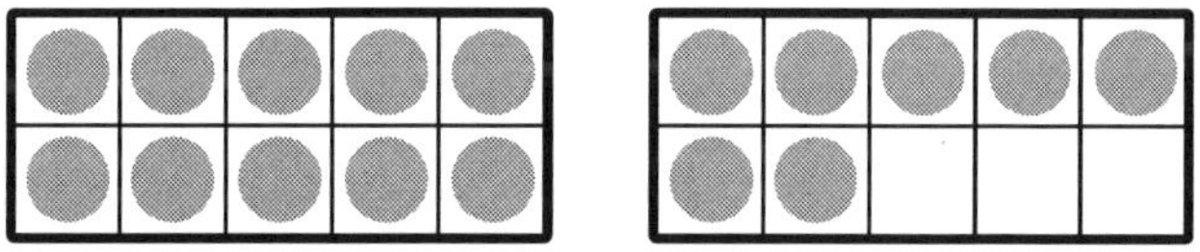

4. Encourage your child to explain what they saw (for example, "I saw 10 and 7 more").
5. Repeat using different flashcards.

This activity helps your child recognise larger numbers by spotting groups within groups, rather than counting every dot.

The Biggest Challenge!

Have ready:

- A flashcard showing 10 dots (a full ten-frame)
- Flashcards showing 1-9
- Number cards 10-19
- Number word cards (eleven–nineteen)

How to play:

1 Place the 10 flashcard on the table so it can be seen.

2 Turn over one dot card and ask:

3. Encourage answers like “ten and six, that makes sixteen!”.

4. Keep going until all cards are turned.

This activity helps your child understand that teen numbers are made from ten and some more (for example, 16 is ten and six).

Answers Explored

Draw it! Number Stories and Patterns

In the 'Draw it!' section, number stories are shown in pairs to help your child spot patterns and connections. Each pair follows a similar structure with small changes.

By looking at what's the same and what's different, children build a stronger understanding of how numbers work – and how they can be used in different ways.

Number Sentence Set	What's the Same?	What's Different?	Why It Matters
6 + 4 = 10 7 + 3 = 10	Both make 10.	Addends (parts) are different.	Encourages children to compare combinations that make the same total and see part-part-whole relationships. Builds mental flexibility with ten frames.
12 - 5 = 7 12 - 6 = 6	Both start with the same number.	Number subtracted.	Shows that subtracting 1 more leads to 1 less—a key insight for mental subtraction. On a ten frame, children can build 12, remove 5, then 6, and compare.
5 + 6 = 11 11 - 6 = 5	Numbers and context match closely.	Focus of question changes from finding the total to working out what remains.	Encourages use of known facts (e.g. 5 + 6 = 11) to solve missing number problems.
8 + 4 = 12 12 - 4 = 8	Same total (12) used.	Addition builds up; subtraction breaks down.	Helps children connect inverse operations—adding and subtracting the same number leads to related results.
(8+5)–3 = 10 (8–3)+5 = 10	Both use the numbers 8, 5 and 3.	The order of ducks flying in and flying away changes.	Helps children see that order matters when mixing addition and subtraction, so they must follow the story carefully.

Group it! Share it! **Multiplication Explored**

Each pair of multiplication number stories in this section changes slightly so that children notice what stays the same and what's different. This helps them understand how changing the order of numbers doesn't change the total.

Number Sentence Set	What's the Same?	What's Different?	Why It Matters
2 x 5 = 10 5 x 2 = 10	- Both stories end with 10 cookies in total. - They both use the numbers 2 and 5.	- In the first story, there are more boxes with fewer cookies. - In the second, there are fewer boxes with more cookies.	This shows your child that the order of numbers (commutativity) in multiplication doesn't change the answer.
3 x 4 = 12 4 x 3 =12	- Both stories have 12 cupcakes total. - They use the same numbers: 3 and 4.	- One story has more rows with fewer cupcakes. The other flips that.	- It helps your child understand that groups can be arranged in different ways but still total the same.
6 x 5 = 30 5 x 6 = 30	- Both have 30 muffins in the end. - The numbers 5 and 6 are used in both.	- One focuses on more trays, the other on more muffins per tray.	This helps children see that multiplication is flexible and still reliable.
5 x 3 = 15 3 x 5 = 15	Both stories use the same numbers (3 and 5) and have the same total (15).	In the first, there are more trays with fewer buns; in the second, fewer trays with more buns.	This shows commutativity — changing the order of numbers in multiplication doesn't change the total.
6 x 4 = 24 4 x 6 = 24	Both use the same numbers (4 and 6) and give the same total (24).	The first shows 4 groups of 6; the second shows 6 groups of 4.	It helps children see that multiplication is flexible — the order of groups doesn't affect the total.

Group it! Share it! Division Explored

The bakery-themed stories in this section explore division through both sharing and grouping structures.
By looking closely at what stays the same and what changes, children begin to understand that sharing is when you know how many groups - but need to find out the size of one group. This differs from grouping where you know the size of the group - but need to find out how many groups altogether.

Number Sentence Set	What's the Same?	What's Different?	Why It Matters
10 ÷ 2 = 5 10 ÷ 2 = 5	Both use 10 cupcakes and the number 2.	One focuses on number of groups (2), result per group (5). The other explores how many groups, not sharing into fixed groups.	One shows equal sharing — dividing into a set number of groups. The other distinguishes between group size and number of groups.
10 ÷ 5 = 2 10 ÷ 5 = 2	Both use 10 doughnuts and the number 5.	One looks at more groups (5), fewer items per group (2). The other is thinking in group size (5) instead of sharing.	One highlights inverse relationship between group number and size. The other shows that division is not just sharing but also making equal groups.
20 ÷ 4 = 5 20 ÷ 4 = 5	Both use 20 buns and the number 5. Total quantity (20) - division.	One focuses on increased quantity and number of groups. The other explores size of group vs. number of groups.	One builds fluency in equal distribution across more groups. The other promotes flexibility in how division is understood.

Number Sentence Set	What's the Same?	What's Different?	Why It Matters
12 ÷ 4 = 3 12 ÷ 4 = 3	Both use 12 muffins and the number 4.	One shares muffins equally between trays; the other groups them into boxes of 4.	Shows that division can mean sharing into equal groups or making groups of a given size.
15 ÷ 3 = 5 15 ÷ 3 = 5	Both use 15 cookies and the number 3.	One shares cookies evenly between trays; the other makes groups of 3.	Reinforces that division can start with the same total but ask for different things — how many in each group or how many groups.

Apply it! Exploring the 4 Operations

The examples in this section help your child talk about patterns and reasoning in all four operations. By looking closely at what stays the same and what changes, children can begin to see patterns and better understand how numbers work. This helps build confidence, mental maths skills and flexible thinking.

Number Sentences	What's the Same?	What's Different?	Why It Matters
12 + 5 = 17 13 + 4 = 17	Both stories have the same total (17).	Addends (parts) are different.	Children can compare how different number combinations create the same total.
18 − 6 = 12 18 − 4 = 14	Both start with the same amount (18).	Different values are subtracted.	Children can see how changing what is taken away changes the result.
4 x 3 = 12 2 x 6 = 12	Both stories result in 12.	Groups and amounts differ.	Children can compare different factor pairs in multiplication.
16 ÷ 4 = 4 16 ÷ 4 = 4	Both use the same total (number of lorries) and involve sharing equally.	One finds how many in each group, the other how many groups.	Recognising both structures deepens children's understanding of division.
(9 + 7) − 3 = 13 (10 + 5) − 2 = 13	Same outcome (13).	Different starting numbers and steps.	It helps children practise reasoning about order and change.

Number Sentences	What's the Same?	What's Different?	Why It Matters
6 + 6 + 2 = 14 (6 + 6) − 2 = 10	Both use the same structure of equal groups.	One adds and the other subtracts.	This helps children notice patterns in repeated addition and subtraction.
7 + 3 + 6 = 16 (7 + 3) − 6 = 4	Both involve adding and taking away groups.	One combines all, the other removes a group.	This shows how the same numbers can tell different operation stories.
4 + 4 + 4 = 12 (4 + 4) − 4 = 4	Both start with repeated groups.	One keeps adding, the other subtracts at the end.	It supports flexible thinking about patterns in multiplication and subtraction.
8 + 2 + 5 = 15 (8 + 2) − 5 = 5	Both include the same sets of numbers.	One adds all, the other takes away.	It builds understanding of addition and subtraction as opposites (the inverse).
5 + 5 = 10 10 − 5 = 5	Both use equal groups and talk about making and losing the same amount.	One doubles, the other halves.	It connects doubling and halving as opposite relationships.

Resources

Counter cut-outs

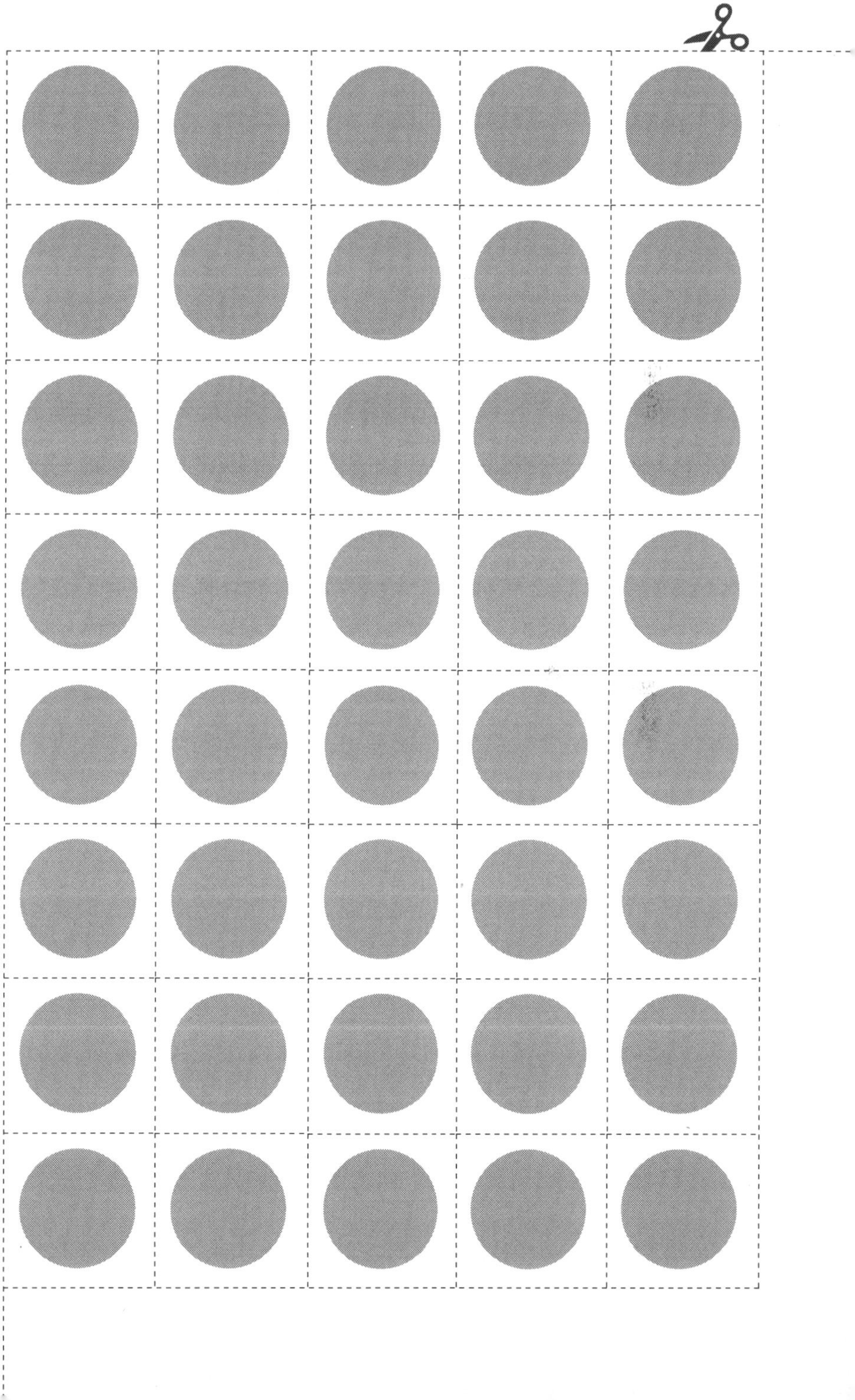

Ten-frame templates

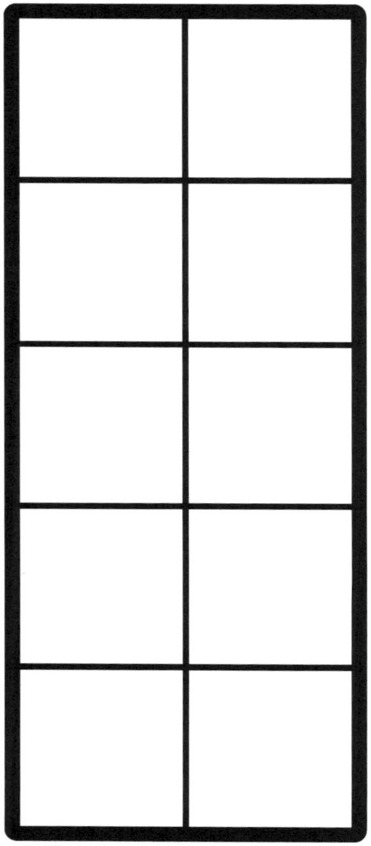

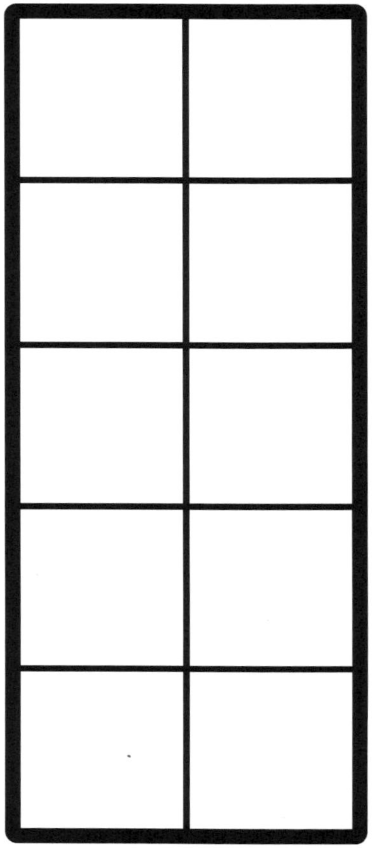

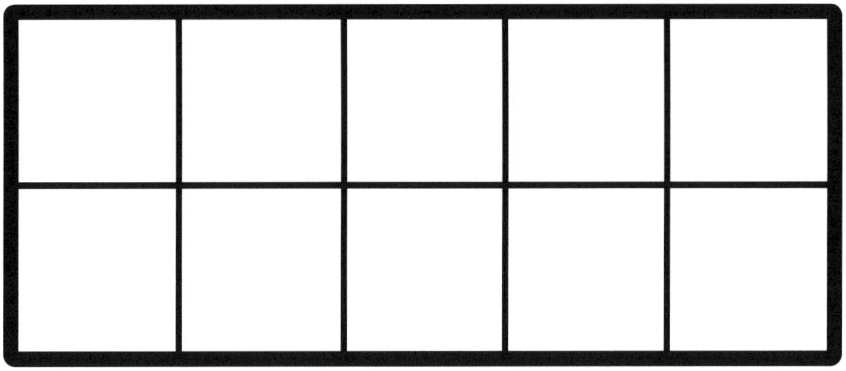

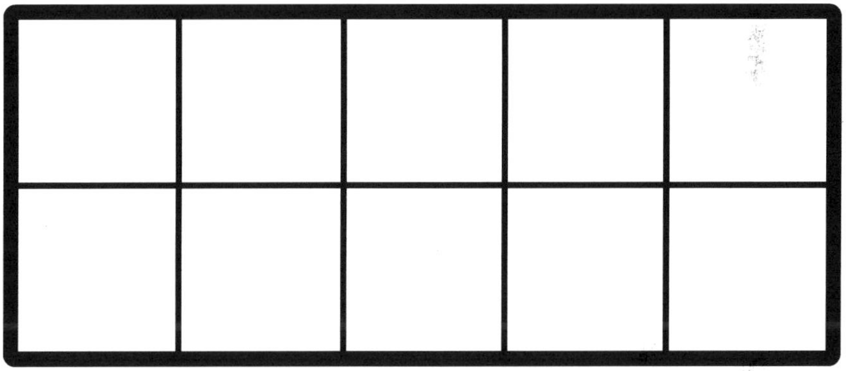

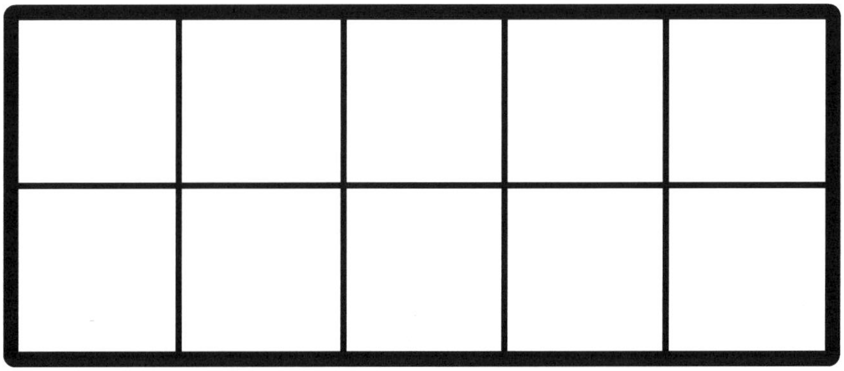

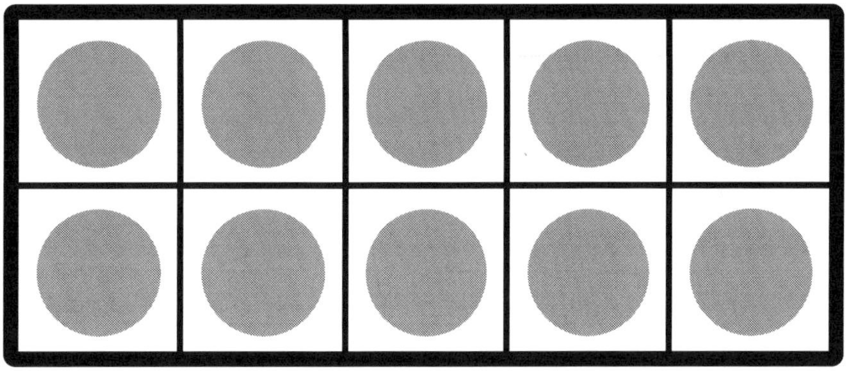

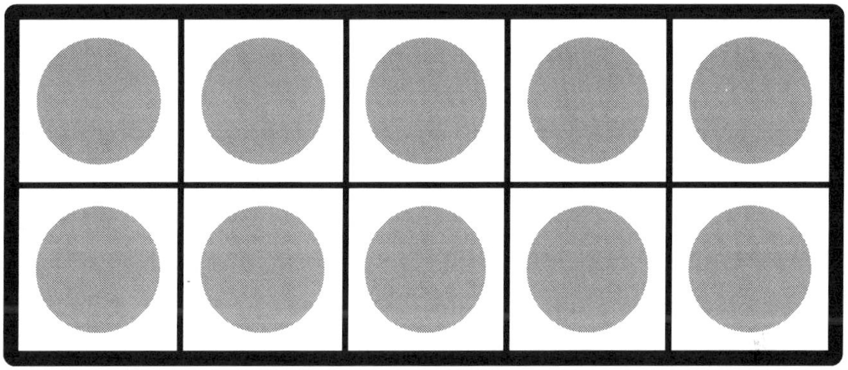

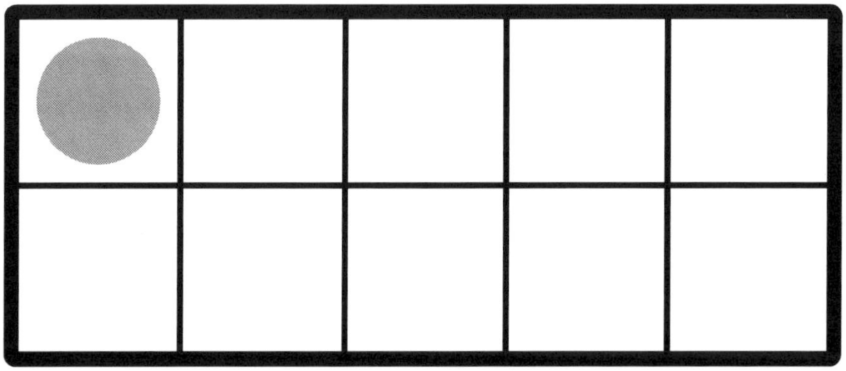

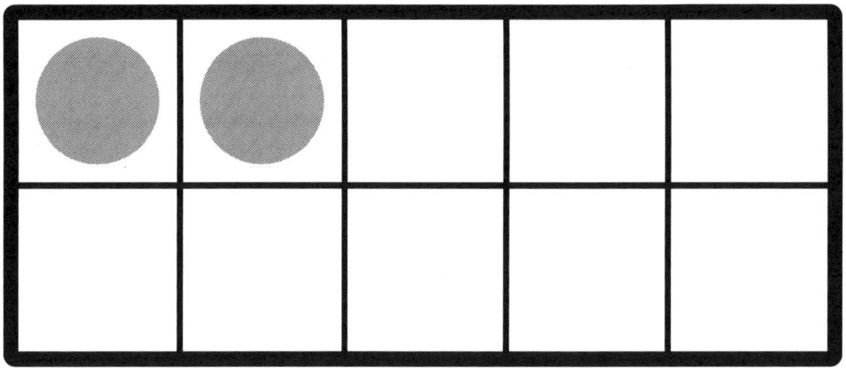

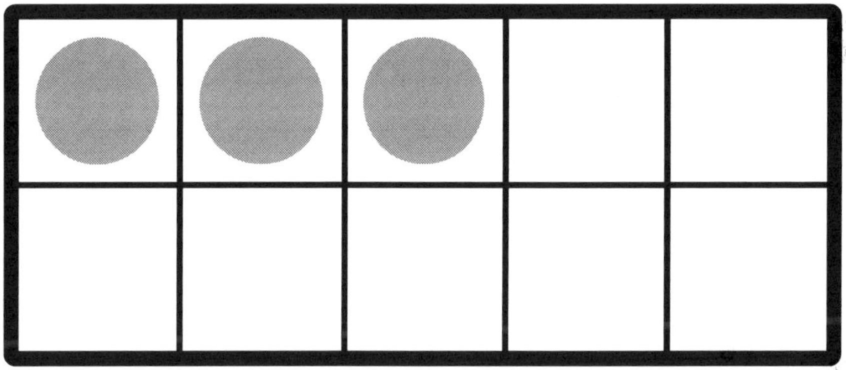

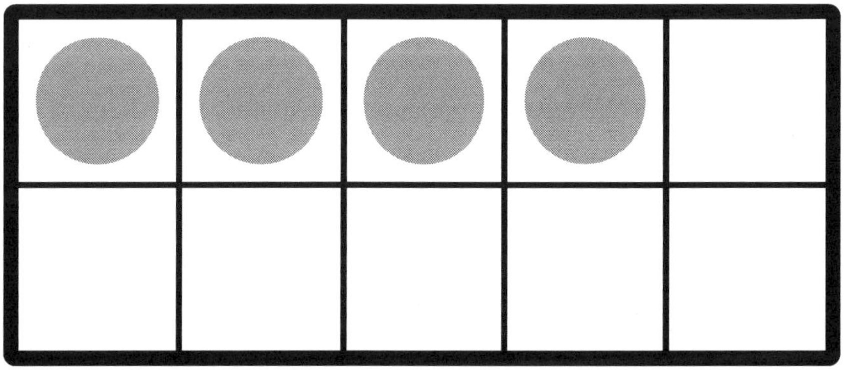

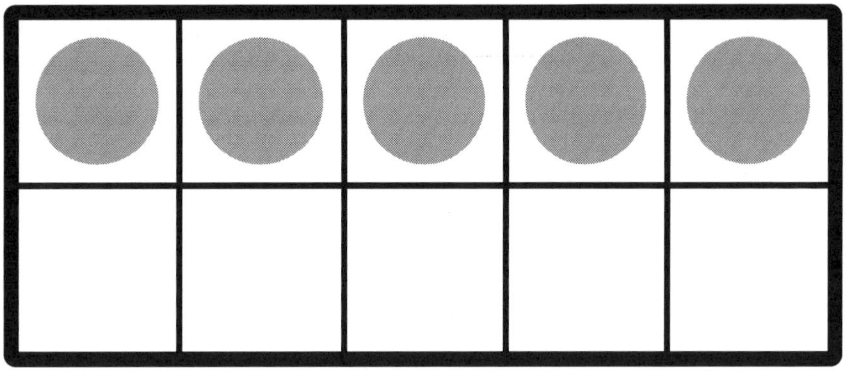

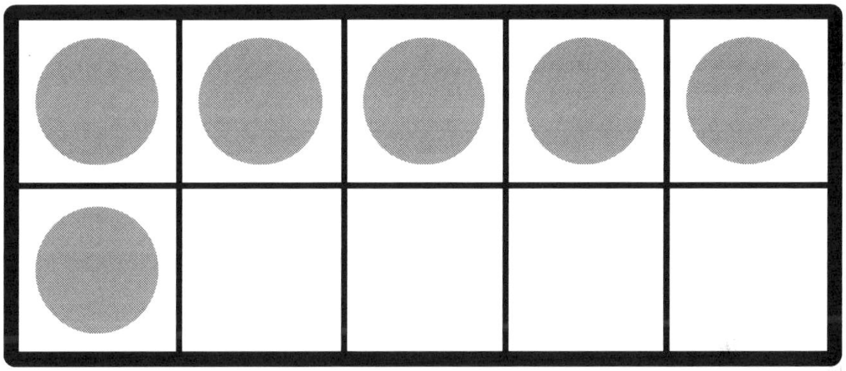

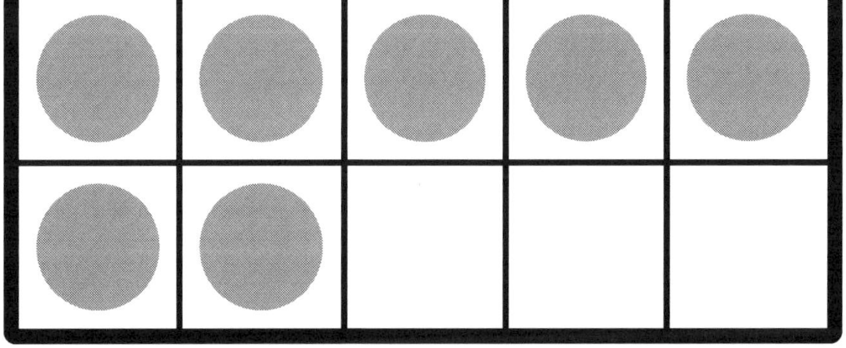

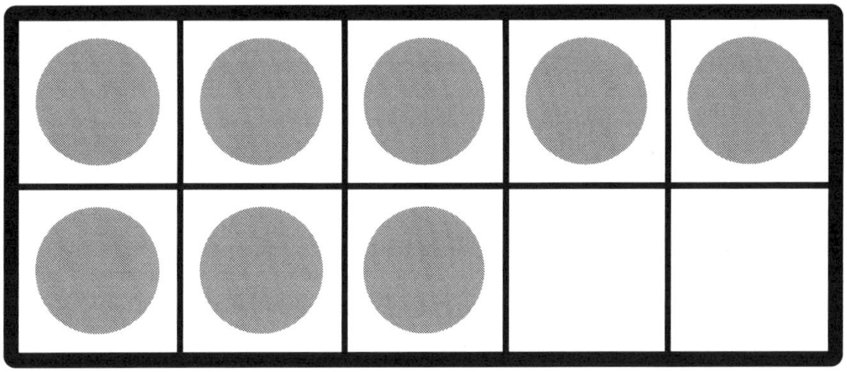

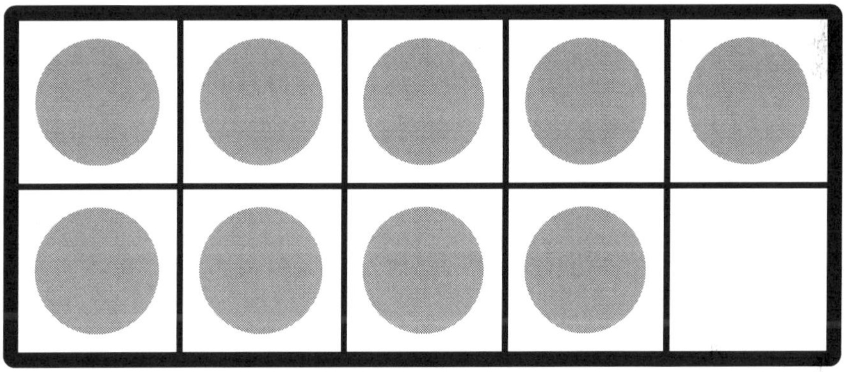

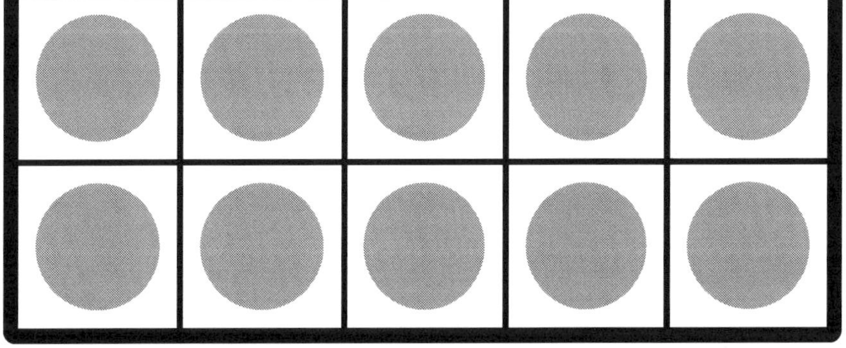

zero	one
two	three
four	five
six	seven
eight	nine
ten	

eleven	twelve
thirteen	fourteen
fifteen	sixteen
seventeen	eighteen
nineteen	twenty
thirty	

5	4	3	2	1

6

7

8

9

13	12	11	10

17	16	15	14

30

20

19

18

Discover, Watch, Play

Discover:

Workshops by Number+

https://www.numberplus.co.uk/

Watch:

NRICH

https://nrich.maths.org/k/

Play:

Interactive Number Frames by Math Learning Centre

https://apps.mathlearningcenter.org/number-frames/

Sponsored by

trudy primary

Scan the QR code

to leave a review

If you found this book valuable, please consider sharing your experience through **an honest rating or review.**

When readers share their thoughts on Amazon, it helps the platform understand who benefits most from this content, connecting the book with more educators just like you.

You can **leave a review** in under a minute by scanning this code with your phone camera.

Made in the USA
Coppell, TX
28 September 2026

91271698R00077